Lecture Notes in Mathematics

A collection of informal reports and seminars
Edited by A. Dold, Heidelberg and B. Eckmann, Zürich

Series: Forschungsinstitut für Mathematik, ETH, Zürich · Adviser: K. Chandrasekharan

24

Joachim Lambek

McGill University, Montreal
Forschungsinstitut fü

Completions of Categories

Seminar lectures given 1966 in Zürich

1966

Springer-Verlag · Berlin · Heidelberg · New York

Library of Congress Catalog Card Number 66-29802. Printed in Germany. Title No. 7344.

Acknowledgement

These notes contain an account of seminar lectures given at the Mathematical Research Institute of the E.T.H. in February 1966, an embryonic version having been presented in a graduate course at McGill in spring 1965.

The author is indebted to McGill University for a generous leave of absence, to the National Research Council of Canada for a Senior Research Fellowship, to the E.T.H. in Zürich for its hospitality, and to Bill Lawvere and Friedrich Ulmer for their stimulation and criticism.

Contents

0. Introduction. We shall call the generalized direct and inverse limits of Kan "supremum" and "infimum" respectively. The derivative terms "sup-complete", "sup-dense", and "sup-preserving" then also have fairly obvious meanings, which will be made precise in the text. One can distinguish these terms from their duals "inf-complete" etc., without being able to tell "left" from "right" or "property" from "co-property".

Can every small category $\underline{A}$ be embedded as a (full) sup-dense subcategory into a sup-complete category $\underline{A}'$? The answer, also noted by others, is yes: Take $\underline{A}'$ to be the category of all functors from $\underline{A}^{o}$, the opposite category of $\underline{A}$, to Ens, the category of sets.

Unfortunately the embedding does not in general preserve sups. However, consider instead the category $\underline{A}''$ of all inf-preserving functors from $\underline{A}^{o}$ to Ens. The embedding of $\underline{A}$ into $\underline{A}''$ is sup-dense and sup-preserving. Moreover $\underline{A}''$ is inf-complete; it is an open problem whether it is also sup-complete.

Luckily there does exist at least one category $\underline{A}'''$ which is sup- and inf-complete with a sup-dense, sup-preserving embedding $\underline{A} \to \underline{A}'''$. To wit, let $\underline{A}'''$ consist of all objects of $\underline{A}''$ which are subobjects in $\underline{A}''$ of products of objects in $\underline{A}$. It is an open problem whether there exists a sup- and inf-complete category $\underline{A}''''$ with a sup- and inf-dense embedding $\underline{A} \to \underline{A}''''$, in analogy to the Dedekind completion of an ordered set.

Now let us drop the assumption that $\underline{A}$ be small. One may still define $\underline{A}'$ and $\underline{A}''$ as before, provided one restricts all functors $T : \underline{A}^{o} \to Ens$ to be "proper". According to Isbell, this means that there exists a small subcategory $\underline{D}$ of $\underline{A}$ such that every element $x \in T(A)$, A in $\underline{A}$, comes from some $y \in T(D)$, D in $\underline{D}$, via some map $f : A \to D$, i.e., such that $x = T(f)(y)$. $\underline{A}$ is still sup-dense in $\underline{A}'$ and $\underline{A}''$, and the embedding $\underline{A} \to \underline{A}''$ is still sup-preserving. While $\underline{A}'$ is sup-complete, $\underline{A}''$ is complete in a different sense: All proper inf-preserving functors to Ens are representable.

How does this "representation completeness" relate to the older forms of completeness? On the one hand, inf-completeness implies representation completeness. This fact, also announced by Benabou, is known to be equivalent to a general Adjoint Functor Theorem. On the other hand, representation completeness implies a kind of sup-completeness. There is also a symmetric theorem which asserts that a modified form of sup-completeness of $\underline{A}$ is equivalent to the representability of certain functors $\underline{A} \to Ens$, and that both conditions on $\underline{A}$ are equivalent to the corresponding conditions on $\underline{A}^{o}$, the opposite category.

We obtain a form of the Special Adjoint Functor Theorem which appears to be slightly more general than any in the literature. (This result is required to show the sup-completeness of $\underline{A}'''$ above.) We also give new sufficient conditions for inf-completeness to imply sup-completeness.

To illustrate completions of small categories, we first consider the example in which $\underline{A}$ is a group G. Then $\underline{A}'' = \underline{A}'$ is the category of all permutational representations of G^{o}.

In another application, we let $\underline{A}$ be a subcategory of an equationally defined category $\underline{C}$ of algebras with finitary operations. If $\underline{A}$ contains a free algebra with sufficiently many generators, then $\underline{A}''$ is equivalent to a subcategory $\underline{A}^{*}$ of $\underline{C}$. $\underline{A}^{*}$ consists of all algebras C such that $\left[-, C\right] : \underline{A}^{o} \to \mathrm{Ens}$ preserves infs. When $\underline{C}$ is the category of all R-modules, this result was first obtained by Ulmer.

Finally, when $\underline{C}$ is the category of all R-modules, under fairly generous conditions on $\underline{A}$, $\underline{A}^{*}$ consists of all R-modules C such that every nonzero submodule of C has a nonzero factor module in $\underline{A}$. Prior to showing this, we make a general study of certain pairs of classes of R-modules, as exemplified by the following pairs of classes of Abelian groups: torsion, torsion-free; divisible, reduced.

It will be assumed that the reader is familiar with what is common to the standard expositions of Category Theory [MacLane, Freyd, Mitchell], in particular with the following concepts: category, functor, natural transformation, monomorphism and epimorphism, subobject and quotient object, pullback and pushout, Yoneda's Lemma [see MacLane, p.54], equivalence of categories, adjoint functors, representable functors.

Subcategory will always mean full subcategory, embedding will always mean a full and faithful functor. Some other well-known concepts will be redefined in Section 1, to allow for some idiosyncracy in terminology.

I have attempted to make these notes readable, at the risk of including some so-called "folk-theorems". For proofs in the literature, the reader is sent to the recent book by Mitchell whenever possible. For some important results however, the papers by Isbell must be consulted.

1. Terminology. It is understood that for every pair of objects A, B of a category $\underline{A}$ there is given a set Hom (A, B) = [A, B] of maps $a : A \to B$. [A, B] is itself the object of the category of sets, called Ens; this may be taken to be any of the universes of Grothendieck [see Gabriel (1962)]. The category $\underline{A}$ is called small if the class of objects is a set, i.e., an object of Ens.

Regrettably, we do not find it convenient to use the styles "$\underline{A}$", "A", "a" for "category", "object" and "map" consistently. Frequently we shall consider categories A with objects a and maps α. Functors will be denoted by capital Roman or Greek letters.

A diagram is the same thing as a functor $\Gamma : I \to \underline{A}$. If this terminology is used, I is called the index category. In particular, with each object A of $\underline{A}$ we may associate the constant diagram $A_I : I \to \underline{A}$, defined by

$$A_I\,(i) = A, \quad A_I\,(\iota) = 1_A,$$

for each object i and map ι of I.

An upper bound (A, u) of a diagram $\Gamma : I \to \underline{A}$ consists of an object A and a natural transformation $u : \Gamma \to A_I$. (Of course, it would have been sufficient to specify u alone.) The upper bound (A, u) of Γ will be called a least upper bound or supremum of Γ if for every upper bound (A', u') of Γ there exists a unique map $a : A \to A'$ such that $au(i) = u'(i)$ for all objects i of I. This situation is illustrated by a commutative diagram:

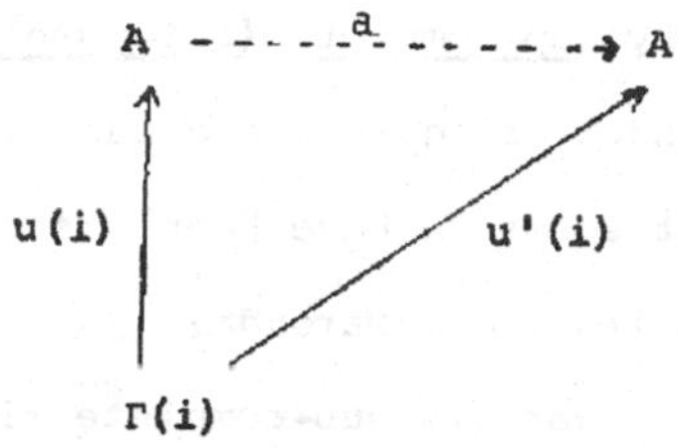

We write $\sup \Gamma = (A, u)$, and sometimes loosely $\sup \Gamma = A$. Actually, the object A is unique only up to isomorphism. We shall use the following result of Kan:

PROPOSITION 1 A. Let $\Gamma : I \to \underline{A}$. Then $\sup \Gamma = A$ if and only if $[\Gamma, A'_I] \cong [A, A']$ is a natural isomorphism in A', where A' is any object of $\underline{A}$.

Dually, one may also introduce the <u>greatest lower bound</u> or <u>infimum</u> of Γ, written $\inf \Gamma = (A, u)$, where $u : A_I \to \Gamma$. The inf in $\underline{A}$ is the same as the sup in $\underline{A}^o$.

The supremum has been called "direct limit", "co-limit", or "right root" by various authors. The present terminology was suggested by Amitsur. While the classical direct limit over a directed set I may be viewed as a special kind of sup, not every sup is a classical direct limit. For example, every Abelian group is a sup of free Abelian groups, but only torsion-free groups are classical direct limits of free ones.

Other special cases of $\sup \Gamma$ are the <u>sum</u> $\sum_{i \in I} \Gamma(i)$, in case I is just a set or small "discrete" category, the <u>cokernel</u> (or "co-equalizer") cok (a, b) of a pair of maps $a, b : A \to A'$, and the <u>pullback</u>. Dually, $\inf \Gamma$ has the <u>product</u> $\Pi_{i \in I} \Gamma(i)$, the <u>kernel</u> ker (a, b), and the <u>pushout</u> as special cases.

We shall call a category <u>sup-complete</u> (<u>inf-complete</u>) if every diagram with small index category has a sup (inf). The following fundamental result seems to have been discovered by Grottendieck, Eckmann and Hilton, and Maranda:

PROPOSITION 1 B. A category is sup-complete (inf-complete) if and only if every set of objects has a sum (product) and every pair of maps between the same objects has a cokernel (kernel).

It is clear that Ens is both sup- and inf-complete. Moreover, if $\underline{A}$ is small, the category of all functors from $\underline{A}$ to Ens is both sup- and inf-complete. We shall put this last observation into a more general context. We shall write $[\underline{A}, \underline{C}]$ for the category of all functors from $\underline{A}$ to $\underline{C}$, it being assumed that $\underline{A}$ is small. The maps of $[\underline{A}, \underline{C}]$ are natural transformations.

PROPOSITION 1 C. If $\underline{C}$ is sup-complete (inf-complete) and $\underline{A}$ is small, then $[\underline{A}, \underline{C}]$ is also sup-complete (inf-complete).

<u>Proof</u> (sketched). Let I be small and consider any diagram $\Gamma : I \to [\underline{A}, \underline{C}]$. Define $\Gamma'(A)(i) = \Gamma(i)(A)$, for all objects A of $\underline{A}$ and i of I. Then $\Gamma'(A) : I \to C$ is a functor. Let sup $\Gamma'(A) = \bigl(F(A), u'(A)\bigr)$, and write $u(i)(A) = u'(A)(i) :$ $\Gamma(i)(A) \to F(A)$. One easily verifies that $F : \underline{A} \to \underline{C}$ is a functor and $u : \Gamma \to F_I$ a natural transformation. Suppose $v : \Gamma \to G_I$ is another natural transformation, where $G : \underline{A} \to \underline{C}$. Put $v'(A)(i) = v(i)(A)$ and verify that $v'(A) : \Gamma'(A) \to G(A)$ is a natural transformation. Then there exists a unique map $f(A) : F(A) \to G(A)$ such that $f(A)u'(A)(i) = v'(A)(i)$, that is to say, $fu(i) = v(i)$. Finally, one verifies that $f : F \to G$

is a natural transformation. It follows that sup $\Gamma = (F, u)$.

This proof is illuminated by the isomorphism of categories $\left[\underline{A},[I, \underline{C}]\right] \cong [I\times\underline{A}, \underline{C}] \cong [\underline{A}\times I, \underline{C}] \cong \left[I, [\underline{A}, \underline{C}]\right]$.
We have presented it here, because the construction of the sup in $[\underline{A}, \underline{C}]$ will be of importance later.

2. Generating and sup-dense subcategories.

With any functor $T : \underline{A} \to Ens$ we shall associate canonically a category X_T and a diagram $\Gamma_T : X_T \to \underline{A}$. For the sake of simplicity, we shall assume that T sends distinct objects of $\underline{A}$ to disjoint sets. This assumption is frequently satisfied; but even if T does not have this property, it is easy to construct a functor $T' \cong T$ which does have it. For example, we might put $T'(A) = \big[[A, -], T\big]$ in $[\underline{A}, Ens]$, then $T' \cong T$, by Yoneda's Lemma, and $A \neq A'$ implies that $[A, -] \neq [A', -]$, hence that $T'(A)$ does not meet $T'(A')$. (The reader will recall that in the usual definition of a category it is assumed that the sets $[A, B]$ are all disjoint.)

The class of objects of X_T is the union of the disjoint sets $T(A)$, A in $\underline{A}$. The maps of X_T are triples $\xi = (x, x', a)$ where $x \in T(A)$ and $x' \in T(A')$ are objects of X_T, and $a : A \to A'$ is a map of X_T such that $x' = T(a)(x)$. We write $\xi : x \to x'$; and, if also $\xi' = (x', x'', a')$, we put $\xi'\xi = (x, x'', a'a)$. The diagram $\Gamma_T : X_T \to \underline{A}$ is defined by stipulating that $\Gamma_T(x)$ is the A such that $x \in T(A)$, and $\Gamma_T(\xi)$ is the a such that $\xi = (x, x', a)$.

Remark. Let $\{0\}$ be a typical one-element set (or the one-element set, according to Lawvere (1964)). We may associate with the element $x \in T(A)$ the map $\hat{x} : \{0\} \to T(A)$ such that $\hat{x}(0) = x$. The condition satisfied by (x, x', a) in X_T may then be represented by the commutative diagram:

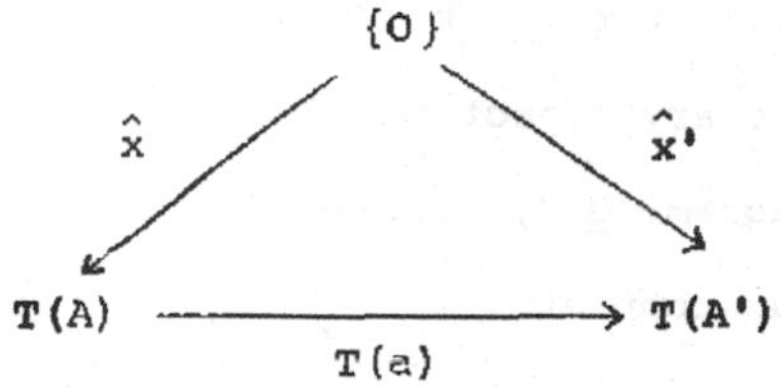

We shall ultimately show that, in some sense, T is the inf of Γ_T. This must be made precise, as T is not an object $\underline{A}$ but only of $[\underline{A}, Ens]$ or, better still, of $[\underline{A}, Ens]^o$, it being assumed that $\underline{A}$ is small. Fortunately, there does exist a canonical embedding $H : \underline{A} \to [\underline{A}, Ens]^o$, where $H(A) = [A, -]$. We shall see that in $[\underline{A}, Ens]^o$, T is in fact the inf of the diagram $H \circ \Gamma_T : X_T \to [\underline{A}, Ens]^o$. However, in the present section, we shall be content with a weaker result concerning Γ_T.

Sometimes we shall consider a functor $T : \underline{A}^o \to Ens$, in which case $\Gamma_T : X_T \to \underline{A}^o$. It is often convenient to view this as a diagram $\Gamma_T^{\,o} : X_T^{\,o} \to \underline{A}$. Consider the canonical embedding $H^o : \underline{A} \to [\underline{A}^o, Ens]$, where $H^o(A) = [\,-, A]$. Then the above mentioned result takes the dual form: T is the sup of the diagram $H^o \circ \Gamma_T^{\,o} : X_T^{\,o} \to \underline{A}$.

Let $\underline{A}$ be a small category and consider a functor $G : \underline{A} \to \underline{B}$. One says that G <u>generates</u> B if for any pair of distinct maps $b_1, b_2 : B \to B'$ in B there exists an object A in $\underline{A}$ and a map $b : G(A) \to B$ such that $b_1 b \neq b_2 b$. This may also be expressed by saying that the functor $G' : B \to [\underline{A}^o, Ens]$ defined by $G'(B) = [G -, B]$ is faithful. In the special case when G is the inclusion of $\underline{A}$ in $\underline{B}$, one calls $\underline{A}$ a <u>generating subcategory</u> of $\underline{B}$. The term "cogenerate" is defined dually.

PROPOSITION 2.1. Let $\underline{A}$ be small, $G : \underline{A} \to \underline{B}$. Then G generates $\underline{B}$ if every object of $\underline{B}$ is a quotient object of a sup of some diagram $\underline{G \circ \Gamma}$, where $\Gamma : I \to \underline{A}$. The converse holds when $\underline{B}$ is sup-complete.

__Proof__. First, assume that every object of $\underline{B}$ has the indicated form. Suppose $b_1, b_2 : B \to B'$, then we have a diagram $\Gamma : I \to A$, with $\sup (G \circ \Gamma) = (B^*, u)$, and an epimorphism $p : B^* \to B$. Suppose $b_1 b = b_2 b$ for all $b : G(A) \to B$, A in $\underline{A}$. Then $b_1 pu(i) = b_2 pu(i)$ for all i in I. It follows from the definition of sup that $b_1 p = b_2 p$, hence $b_1 = b_2$. Thus G generated $\underline{B}$.

Conversely, assume that G generates $\underline{B}$, and also that $\underline{B}$ is sup-complete. With any object B in $\underline{B}$ there is associated the functor $T = [G -, B] : \underline{A}^o \to \underline{B}$, hence the small category $X = X_T{}^o$ and the diagram $\Gamma = \Gamma_T{}^o : X \to \underline{A}$. The class of objects of X is the union of all $[A, B]$, A in $\underline{A}$, and the maps of X are the triples $\xi = (x, x', a)$ such that $x' = [G(a), B](x) = x\, G(a)$. Moreover, $\Gamma(x)$ is the A such that $x : G(A) \to B$ (we assume, for simplicity's sake that the sets $[G(A), B]$ are all disjoint), and $\Gamma(\xi) = a$. For any object $x \in [G(A), B]$, we put $1(x) = x : G\big(\Gamma(x)\big) \to B$, and it is easily verified that $1 : G \circ \Gamma \to B_X$ is a natural transformation.

Now let $\sup(G \circ \Gamma) = (B^*, u)$, then there exists a unique map $p : B^* \to B$ such that $pu(x) = x$. Moreover p is epi, since $b_1, b_2 : B \to B'$ and $b_1 p = b_2 p$ would yield $b_1 x = b_2 x$ for all x in X, hence $b_1 = b_2$. Thus B is a quotient object of sup $(G\circ\Gamma)$, and the proof is complete.

Again consider a functor $G : \underline{A} \to \underline{B}$ where $\underline{A}$ is small. Following Isbell (1960), we call G left adequate for $\underline{B}$ if the canonically associated functor $G' : \underline{B} \to [\underline{A}^o, \text{Ens}]$ is an embedding, i.e., faithful and full. Thus, if G is left adequate for $\underline{B}$, then G generates $\underline{B}$. We shall call G sup-dense if every object in $\underline{B}$ is the sup of some diagram $G\circ\Gamma$, where $\Gamma : I \to \underline{A}$. When G is the inclusion of $\underline{A}$ in $\underline{B}$, we call $\underline{A}$ a left adequate or sup-dense subcategory of $\underline{B}$, respectively. As had been noted by Isbell, the canonical embedding $H^o : \underline{A} \to [\underline{A}^o, \text{Ens}]$ is left adequate. Right adequate and inf-dense functors are defined dually. For example, the canonical embedding $H : \underline{A} \to [\underline{A}, \text{Ens}]^o$ is right adequate. The following has also been observed by Ulmer:

COROLLARY. If $\underline{A}$ is small, $\underline{B}$ is sup-complete, and $G : \underline{A} \to \underline{B}$ is left adequate, then G is sup-dense. In particular, every functor $T : \underline{A}^o \to \text{Ens}$ is the sup of representable functors $[-, A]$, A in $\underline{A}$.

Actually, this result holds even without the assumption that $\underline{B}$ is sup-complete and will be proved in greater generality in Section 5.

Proof. In view of Proposition 2.1 , or rather its proof, we have $\Gamma : I \to \underline{A}$, $\sup(G\circ\Gamma) = (B^*, u)$, and an epimorphism $p : B^* \to B$ such that $pu(x) = x$.

Let $t(A) : [G(A), B] \to [G(A), B^*]$ be defined by $t(A)(x) = u(x)$. It is easily verified that this is natural in A, hence we have a natural transformation $t : [G -, B] \to [G -, B^*]$.

Now the functor $G' : \underline{B} \to [\underline{A}^{o}, \text{Ens}]$ is assumed to be full, hence the mapping $[B, B^*] \to \big[[G -, B], [G -, B^*]\big]$ is onto. Therefore there exists a map $b : B \to B^*$ such that $[G -, b] = t$, that is to say

$$u(x) = t(A)(x) = [G(A), b](x) = bx,$$

for any x in X (see the proof of Proposition 2.1). Thus $bpu(x) = bx = u(x)$, and so $bp = 1$, by definition of sup. Therefore $pbp = p$, hence also $bp = 1$, since p is epi. Thus $\sup \Gamma = (B, 1)$, where $1(x) = x$, and the proof is complete.

As we have seen, the canonical functor $H^{o} : \underline{A} \to [\underline{A}^{o}, \text{Ens}]$ is a sup-dense embedding into a sup-complete category, hence dually the functor $H : \underline{A} \to [\underline{A}, \text{Ens}]^{o}$ is an inf-dense embedding into an inf-complete category. One would be tempted to call $[\underline{A}^{o}, \text{Ens}]$ the "sup-completion" of $\underline{A}$, were it not for the fact that H^{o} does not in general preserve sups. While many examples will be discussed systematically in Sections 9 and 10, we mention now a simple example due to Ulmer (1966): The Abelian group of rationals is a classical direct limit of subgroups isomorphic to the group of integers (all fractions with denominator n), but the same directed set of groups has direct limit zero in the category of free Abelian groups.

We saw that any functor $G : \underline{A} \to \underline{B}$ has canonically associated with it a functor $G' : \underline{B} \to [\underline{A}^{o}, \text{Ens}]$, where $G'(B)(A) = [G(A), B]$. An easy computation shows that $G' \circ G \cong H^{o}$.

In particular, there is a functor $H' : [\underline{A}, \mathrm{Ens}]^{o} \to [\underline{A}^{o}, \mathrm{Ens}]$ such that $H' \circ H \cong H^{o}$. Dually there is a functor $H^{o\prime} : [\underline{A}^{o}, \mathrm{Ens}] \to [\underline{A}, \mathrm{Ens}]^{o}$ such that $H^{o\prime} \circ H^{o} \cong H$. It is not difficult to verify that $H^{o\prime}$ is the left adjoint of H'.

We shall abbreviate $H'(T) = T^{*}$, $H^{o\prime}(U) = U^{+}$. There exist canonical maps $T^{*} \to T^{*+*}$ and $T^{*+*} \to T^{*}$ such that their composite $T^{*} \to T^{*}$ is the identity. Unfortunately, this is not enough to establish an equivalence between the category of all T^{*} and the category of all U^{+}. As Isbell (1960) pointed out, the functors $*$ and $+$ do induce an equivalence between the "reflexive" subcategories of all T in $[\underline{A}, \mathrm{Ens}]^{o}$ such that $T^{*+} \cong T$ and all U in $[\underline{A}^{o}, \mathrm{Ens}]$ such that $U^{+*} \cong U$. In this way, one can construct a kind of intersection of $[\underline{A}^{o}, \mathrm{Ens}]$ and $[\underline{A}, \mathrm{Ens}]^{o}$. Unfortunately, this intersection is rather small. When $\underline{A}$ is a group G with more than two elements, Isbell (1964) showed that this intersection has just three objects, the regular permutational representation of G^{o} and two other trivial functors, hence is far from complete (see Section 8 below).

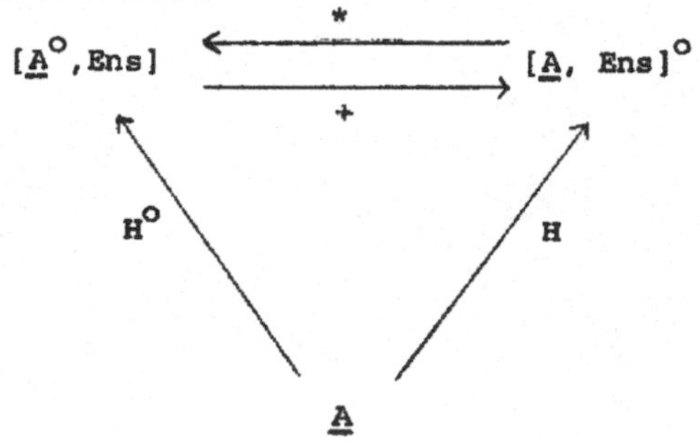

The above situation admits a generalization.
Let $\Phi : \underline{A}^o \times \underline{B} \rightarrow Ens$ be a given bifunctor, where $\underline{A}$ and $\underline{B}$ are given small categories. This determines canonically functors $F : \underline{A} \rightarrow [\underline{B}, Ens]^o$ and $G : \underline{B} \rightarrow [\underline{A}^o, Ens]$ defined by

$$F(A)(B) = \Phi(A,B) = G(B)(A).$$

These, in turn, give rise to functors G' and F' such that $G' \circ G = H$ and $F' \circ F = H^o$. It may be verified that G' is the left adjoint of F'.

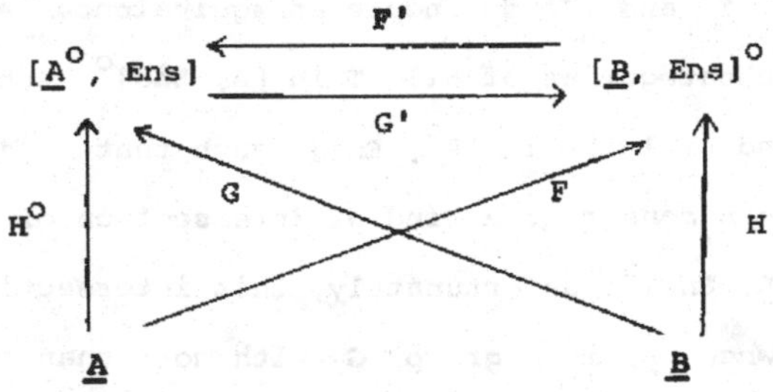

That the original situation is indeed a special case of this is seen by considering the bifunctor $Hom : \underline{A}^o \times \underline{A} \rightarrow Ens$.

3. Limit preserving functors.

We say that the functor $F : \underline{A} \to \underline{B}$ preserves sups if, for every diagram $\Gamma : I \to \underline{A}$,

$$\sup \Gamma = (A, u) \Rightarrow \sup (F \circ \Gamma) = \big(F(A), F \circ u\big)$$

Here the small circle denotes composition of concrete mappings:

$$(F \circ \Gamma)(i) = F\big(\Gamma(i)\big), \quad (F \circ u)(i) = F\big(u(i)\big),$$

for any object i in I. The definition does not presume completeness of $\underline{A}$, hence we cannot, in general, apply Proposition 1B to replace preservation of sups by preservation of sums and cokernels. Preservation of infs is defined dually. The following result and its corollary (due to Eckmann and Hilton) are well-known:

PROPOSITION 3 A. For any object A of $\underline{A}$, the functors $[A, -] : \underline{A} \to \mathrm{Ens}$ and $[-, A] : \underline{A}^{\circ} \to \mathrm{Ens}$ preserve infs.

COROLLARY. If $F : \underline{A} \to \underline{B}$ is left adjoint to $G : \underline{B} \to \underline{A}$, then F preserves sups and G preserves infs.

PROPOSITION 3.1. The functor $F : \underline{A} \to \underline{B}$ preserves sups if and only if $[F -, B] : \underline{A}^{\circ} \to \mathrm{Ens}$ preserves infs for all B in $\underline{B}$.

<u>Proof</u>. First, assume that F preserves sups. The functor $[F -, B]$ arises by composition from the inf-preserving functors $[-, B] : \underline{B}^{o} \to Ens$ and $F^{o} : \underline{A}^{o} \to \underline{B}^{o}$, hence also preserves infs.

Conversely, assume that $[F -, B]$ preserves infs for all B in $\underline{B}$. Let $\Gamma : I \to \underline{A}$ and suppose sup $\Gamma = (A, u)$. Consider the diagram $i \rightsquigarrow [F(\Gamma(i)), B]$, i in I. Its inf is $([F(A), B], v)$, where $v(i) = [F(u(i)), B]$. We claim that sup $(F \circ \Gamma) = (F(A), F \circ u)$.

Indeed, let $t(i) : F(\Gamma(i)) \to B$ be natural in i. As in the remark in Section 2, we associate with this a mapping $\hat{t}(i) : \{0\} \to [F(\Gamma(i)), B]$. Hence there exists a unique map $f : \{0\} \to [F(A), B]$ such that $[F(u(i)), B]f = \hat{t}(i)$. Writing $f = \hat{a}$, where $a = f(0)$, we have a unique $a \in [F(A), B]$ such that $aF(u(i)) = t(i)$, as required.

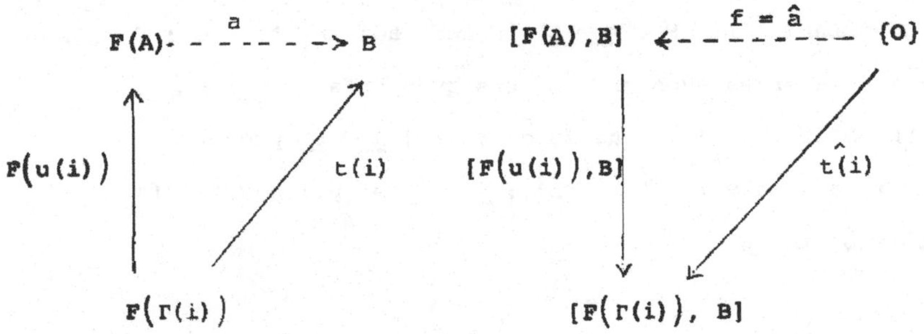

LEMMA 3.1. Let T and T' be isomorphic functors from $\underline{A}$ to $\underline{B}$. Then T preserves infs if and only if T' does.

The proof is routine and will be omitted.

PROPOSITION 3.2. Given an embedding $F : \underline{A} \to \underline{C}$, let $\underline{B}$ consist of all B in $\underline{C}$ such that $[F -, B] : \underline{A}^{o} \to \mathrm{Ens}$ preserves infs. Then

a) $\underline{B}$ contains $F(\underline{A})$, the image of F;

b) $\underline{B}$ is the largest subcategory of $\underline{C}$ such that the induced embedding $\underline{A} \to F(\underline{A}) \to \underline{B}$ preserves sups;

c) $\underline{B}$ is closed under infs in $\underline{C}$.

Proof. (a) Since F is faithful and full, $[F -, F(A)] \cong [-, A]$. The latter functor preserves infs, by Proposition 3 A, hence so does the former , by the above Lemma.

(b) This follows immediately from Proposition 3.1.

(c) Consider any diagram $\Delta : J \to \underline{B}$ with $\inf \Delta = (C, v)$ in $\underline{C}$. We want to show that C is in $\underline{B}$. Assume $\Gamma : I \to \underline{A}$ with $\sup \Gamma = (A, u)$, then $\sup (F \circ \Gamma) = \big(F(A), F \circ u\big)$ in $\underline{B}$. We claim that this remains true in $\underline{B} \cup \{C\}$, so that C must be in $\underline{B}$, by (b).

Let $t(i) : F\big(\Gamma(i)\big) \to C$ be natural in i, where i is any object of I. For each j in J, we have $v(j)t(i) : F\big(\Gamma(i)\big) \to \Delta(j)$ in $\underline{B}$. Therefore there exists a unique $x(j) : F(A) \to \Delta(j)$ such that $v(j)t(i) = x(j)F\big(u(i)\big)$. It may easily be shown that $x(j)$ is natural in j. Hence there exists an unique $y : F(A) \to C$ such that $v(j)y = x(j)$.

Therefore

$$v(j)t(i) = x(j)F\big(u(i)\big) = v(j)y\,F\big(u(i)\big),$$

hence $t(i) = y\,F\big(u(i)\big)$, One easily verifies that y is unique with this property, and this completes the proof.

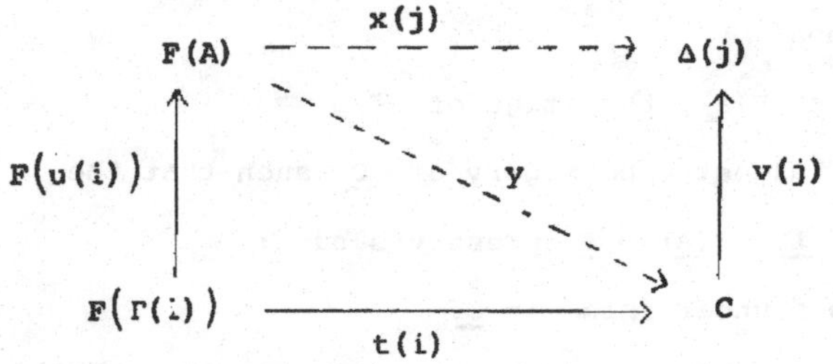

<u>COROLLARY</u>. Every inf-dense embedding preserves sups. In particular, the canonical embedding $H : \underline{A} \to [\underline{A}, \text{Ens}]^{o}$ preserves sups and, dually, the canonical embedding $H^{o} : \underline{A} \to [\underline{A}^{o}, \text{Ens}]$ preserves infs.

<u>Proof</u>. Suppose $F : \underline{A} \to \underline{C}$ is an inf-dense embedding. Let $\underline{B}$ be constructed as above. Now $\underline{B}$ contains $F(\underline{A})$ and for any object C of $\underline{C}$ there is a diagram $\Gamma : I \to \underline{A}$ such that $C = \inf\,(F \cdot \Gamma)$. It follows from the proposition that C is in $\underline{B}$. Thus $\underline{C} = \underline{B}$.

The rest follows from the corollary to Proposition 2.1.

Let $\underline{A}$ be any small category. We shall write $[\underline{A}, \text{Ens}]_{\text{inf}}$ for the category of all inf-preserving functors from $\underline{A}$ to Ens, a subcategory of $[\underline{A}, \text{Ens}]$. In view of Proposition 3A, the canonical embeddings $H^{o} : \underline{A} \to [\underline{A}^{o}, \text{Ens}]$ and $H : \underline{A} \to [\underline{A}, \text{Ens}]^{o}$ induce embeddings of $\underline{A}$ into $[\underline{A}^{o}, \text{Ens}]_{\text{inf}}$ and $[\underline{A}, \text{Ens}]_{\text{inf}}{}^{o}$

respectively. We shall often denote these induced embeddings by H^{o} and H also.

PROPOSITION 3.3. Given a small category $\underline{A}$, the canonical functor $\underline{A} \to [\underline{A}^{o}, \mathrm{Ens}]_{\mathrm{inf}}$ is a sup-dense, sup-preserving, and inf-preserving embedding into an inf-complete category.

We refrain from spelling out the dual statement.

Proof. That the embedding is sup-dense follows from the corollary to Proposition 2.1. That it is inf-preserving then follows from the corollary to Proposition 3.2. That it preserves sups will follow from Proposition 3.1. if we show that $[\underline{A}^{o}, \mathrm{Ens}]_{\mathrm{inf}}$ consists of all functors $T : \underline{A}^{o} \to \mathrm{Ens}$ such that $[H^{o}-, T] : \underline{A}^{o} \to \mathrm{Ens}$ preserves infs. That $[\underline{A}^{o}, \mathrm{Ens}]_{\mathrm{inf}}$ is inf-complete will follow from Proposition 3.2. and 1C for the same reason.

Now, by Yoneda's Lemma, $[H^{o}-, T] \cong T$. By Lemma 3.1, $[H^{o}-, T]$ preserves infs if and only if T does. This completes the proof.

Unfortunately I do not know whether $[\underline{A}^{o}, \mathrm{Ens}]_{\mathrm{inf}}$ is always sup-complete. It is sup-complete in many examples (see Sections 8 and 10). but is not known to be so in general. However, when it is sup-complete, then it is actually a left reflective subcategory of $[\underline{A}^{o}, \mathrm{Ens}]$, as we shall see.

A subcategory $\underline{B}$ of $\underline{C}$ is said to be left reflective if the inclusion functor $\underline{B} \to \underline{C}$ has a left adjoint. This is equivalent to saying that every object C of $\underline{C}$ has a best approximation in $\underline{B}$, i.e., a map $p : C \to B$ with B in $\underline{B}$ such that, for every map $f : C \to B'$ with B' in $\underline{B}$, there exists a unique map $b : B \to B'$ such that $f = bp$. Under fairly mild assumptions it turns out that p must be an epimorphism: One need only assume that $\underline{B}$ is closed under subobjects in $\underline{C}$ and that every map of $\underline{C}$ has the form me, where m is mono and e is epi.

LEMMA 3.2. Let $\underline{B}$ be a subcategory of $\underline{C}$, and assume that the diagram $\Gamma : I \to \underline{B}$ has $\sup \Gamma = (B, u)$ in $\underline{B}$ and $\sup \Gamma = (C, v)$ in $\underline{C}$. Then the unique map $p : C \to B$ such that $pv(i) = u(i)$ is a best approximation of C in $\underline{B}$.

<u>Proof</u>. Suppose $f : C \to B'$, where B' is in $\underline{B}$. Then $fv(i) : \Gamma(i) \to B'$ is natural in i, for i in I. Hence there exists a unique $b : B \to B'$ such that $bu(i) = fv(i)$, i.e., $bpv(i) = fv(i)$, i.e., $bp = f$.

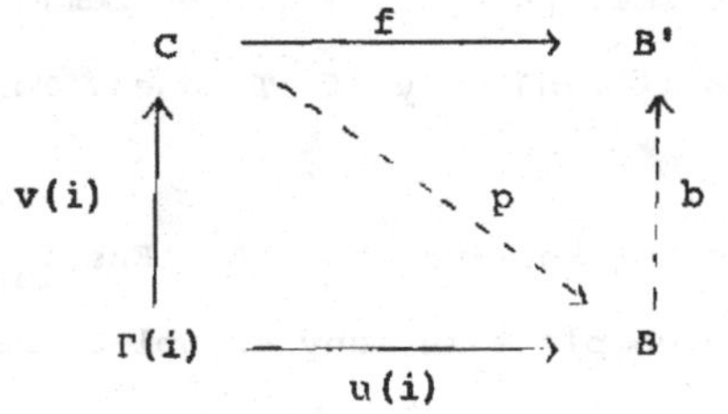

The reader will recall from Section 2 that a functor $F : \underline{B} \to \underline{C}$ was called <u>sup-dense</u> if every object of $\underline{C}$ is the sup of some diagram $F \circ \Gamma$, where $\Gamma : I \to \underline{B}$. It was not then stipulated that I must be small. We now find it convenient to call F <u>properly sup-dense</u> if the same is true with small index category I. Looking again at the corollary to Proposition 2.1, we observe that the result remains valid if "sup-dense" is replaced by "properly sup-dense".

PROPOSITION 3.4. Assume that $\underline{B}$ is a properly sup-dense subcategory of $\underline{C}$ and that $\underline{B}$ is sup-complete, then $\underline{B}$ is a left reflective subcategory. In the converse direction, if $\underline{B}$ is a left reflective subcategory of a sup-complete category $\underline{C}$, then $\underline{B}$ is sup-complete. In particular, for any small category $\underline{A}$, $[\underline{A}^o, \text{Ens}]_{inf}$ is sup-complete if and only if it is a left reflective subcategory of $[\underline{A}^o, \text{Ens}]$.

Proof. Let C be any object of $\underline{C}$. By assumption, there exists a diagram $\Gamma : I \to \underline{B}$ and a natural transformation v such that $\sup\Gamma = (C,v)$ in $\underline{C}$, where I is small. If $\underline{B}$ is sup-complete, then also $\sup\Gamma = (B,u)$ in $\underline{B}$. By the lemma, C has a best approximation in $\underline{B}$, thus $\underline{B}$ is left reflective.

Conversely, we assume that $\underline{C}$ is sup-complete and $\underline{B}$ is left reflective in $\underline{C}$. Consider any diagram $\Gamma : I \to B$ with small I and let $\sup\Gamma = (C,v)$ in $\underline{C}$. Let $p : C \to B$ be the best approximation of C in $\underline{B}$, then also $\sup \Gamma = (B, p_I \circ v)$ in $\underline{B}$. [See Mitchell, p. 129]

Finally, $[\underline{A}^o, \text{Ens}]_{inf}$ is properly sup-dense in $[\underline{A}^o, \text{Ens}]$, because $H^o(\underline{A})$ is, by the corollary to Proposition 2.1.

Actually, a slightly stronger result than the first statement in Proposition 3.4 may be proved by the same method:

PROPOSITION 3.4'. Assume that $F : \underline{B} \to \underline{C}$ is a proper sup-dense embedding and that $\underline{B}$ is sup-complete. Then F has a left adjoint $G : \underline{C} \to \underline{B}$.

4. A sup-complete sup-dense, sup-preserving extension.

LEMMA 4.1. Given a subcategory $\underline{A}$ of $\underline{C}$, let $\underline{B}$ be the subcategory of $\underline{C}$ whose objects are subobjects of products of objects from $\underline{A}$. Then $\underline{B}$ is closed under products.

The proof is routine and will be omitted.

Let there be given a small category $\underline{A}$. We recall the canonical embedding $H^{o} : \underline{A} \to [\underline{A}^{o}, \text{Ens}]_{inf}$ of $\underline{A}$ into an inf-complete category (Proposition 3.3). Now let $\underline{B}$ be the subcategory of $[\underline{A}^{o}, \text{Ens}]_{inf}$ which consists of all subobjects of products of representable functors $H^{o}A = [\,-, A]$, where A is in $\underline{A}$. Obviously, $\underline{B}$ is closed under subobjects, hence under kernels. Moreover, by the lemma, it is closed under products. Since $[\underline{A}^{o}, \text{Ens}]_{inf}$ is inf-complete, the assumptions of Proposition 1B are satisfied, hence $\underline{B}$ is also inf-complete. (Actually, a closer examination of the argument shows that $\underline{B}$ is closed under infs with small index categories in $[\underline{A}^{o}, \text{Ens}]_{inf}$.

By Proposition 2.1, or rather its dual, we see that H^{o} cogenerates $\underline{B}$, hence $H^{o}(\underline{A})$ is a cogenerating subcategory of $\underline{B}$. Moreover, any object of $\underline{B}$ has a representative set of subobjects, as we shall verify presently. We may therefore apply the Special Adjoint Functor Theorem (Proposition 7.1 below) and deduce that the inclusion functor $\underline{B} \to [\underline{A}^{o}, \text{Ens}]$ has a left adjoint, i.e., $\underline{B}$ is a left reflective subcategory of $[\underline{A}^{o}, \text{Ens}]$. Since the latter category is sup-complete (Proposition 1C), so is $\underline{B}$ (see Proposition 3.4). In view of Proposition 3.3, we thus have:

PROPOSITION 4.1. Given a small category $\underline{A}$, let $\underline{B}$ be the subcategory of $[\underline{A}^{o}, \text{Ens}]_{\text{inf}}$ which consists of all subobjects of products of functors $[-, A]$, A in $\underline{A}$. Then $A \rightsquigarrow [-, A]$ is a sup-dense, inf-preserving and sup-preserving embedding of $\underline{A}$ into the inf-and sup-complete category $\underline{B}$.

We refrain from spelling out the dual of this result.

Proof. It remains to show that any object B of $\underline{B}$ has a representative set of subobjects. Now $H^{o}(\underline{A})$ is a sup-dense subcategory of $[\underline{A}^{o}, \text{Ens}]$ (Corollary to Proposition 2.1), hence so is $\underline{B}$. Therefore the inclusion functor $\underline{B} \to [\underline{A}^{o}, \text{Ens}]$ preserves infs, by the corollary to Proposition 3.2, or rather its dual. In particular, any pullback in $\underline{B}$ remains a pullback in $[\underline{A}^{o}, \text{Ens}]$.

Consider any map $b : B' \to B$ in $\underline{B}$. This is mono if and only if in the pullback

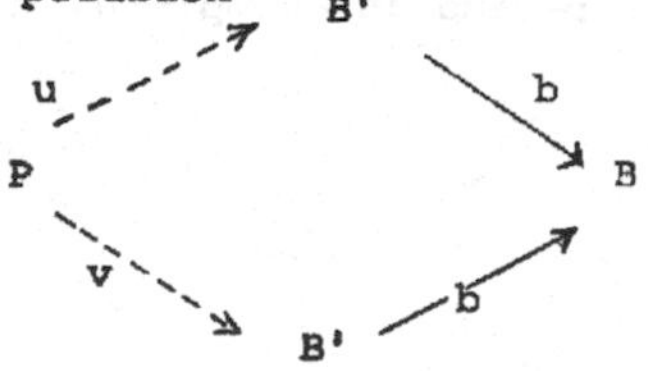

in $\underline{B}$ we have $u = v$. Since this remains a pullback in $[\underline{A}^{o}, \text{Ens}]$, we see that b mono in $\underline{B}$ implies b mono in $[\underline{A}^{o}, \text{Ens}]$. Moreover, it follows from the proof of Proposition 1C that, for each object A of $\underline{A}$,

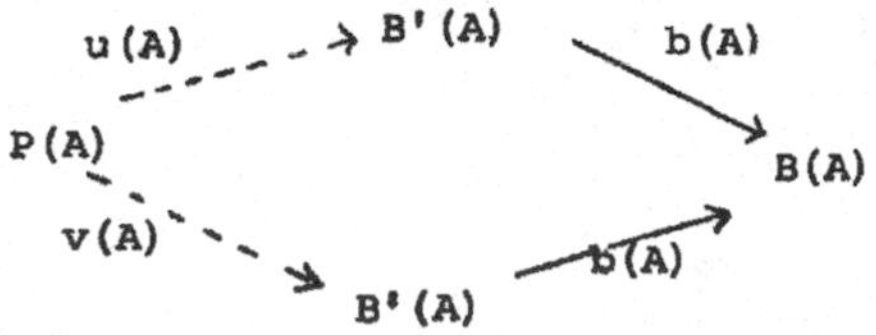

is a pullback in Ens.

Now assume that b is mono in $\underline{B}$. Then, for each A in $\underline{A}$, $b(A) : B'(A) \to B(A)$ is a monomorphism. It gives rise to an isomorphism $t(A) : B'(A) \to B^*(A)$, where $B^*(A) = b(A)(B'(A))$ is a subset of $B(A)$. For any map $a : A \to A'$ in $\underline{A}$, define $B^*(a) = t(A)B'(A)t(A')^{-1}$. Then $B^* : \underline{A}^o \to Ens$ is a functor, and $t : B' \to B^*$ is a natural isomorphism. Thus every subobject $b : B' \to B$ is isomorphic to a special subobject $B^* \to B$, where $B^*(A) \in B(A)$, for each in $\underline{A}$. As the B^* clearly form a set, our proof is complete.

We remark that the embedding $\underline{A} \to \underline{B}$, in addition to being sup-dense, is cogenerating, which property falls just short of inf-denseness. I do not yet know whether for every small category $\underline{A}$ there exists a sup- and inf-dense embedding $\underline{A} \to \underline{C}$, such that $\underline{C}$ is sup- and inf-complete.

5. The completion when $\underline{A}$ is not small.

We now abandon the assumption that $\underline{A}$ is small. The main obstacle in forming $[\underline{A}, \mathrm{Ens}]$ is that $[T, T']$ need not be small if we admit all functors T and T'.

Following Isbell (1960), we call $T : \underline{A} \to \mathrm{Ens}$ <u>proper</u> if there exists a small subcategory $\underline{D}$ of $\underline{A}$, called a <u>dominating set</u> for T, with this property: For all A in $\underline{A}$ and $x \in T(A)$ there exist D in $\underline{D}$, $y \in T(D)$, and $f : D \to A$ such that $x = T(f)(y)$. Letting $\hat{x} : \{0\} \to T(A)$ such that $\hat{x}(0) = x$ (see the Remark in Section 2), we may illustrate this property by a commutative diagram:

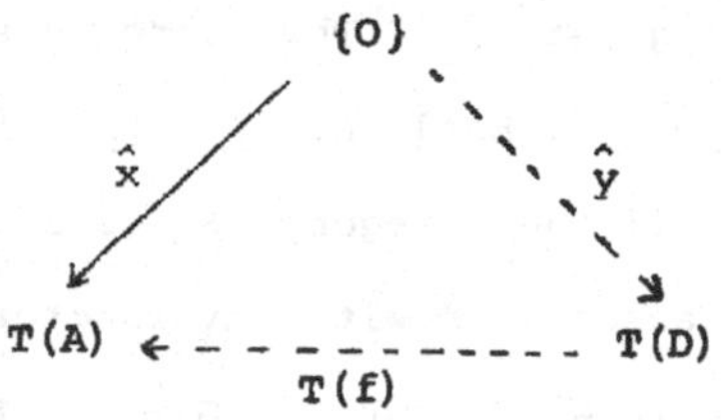

Let $T : \underline{A} \to \mathrm{Ens}$ be a proper functor with dominating set $\underline{D}$. As an object function, T is determined by its restriction $T/_{\underline{D}}$. As a function on maps, it is determined by its value for all maps $D \to A$, where D is in $\underline{D}$ and A in $\underline{A}$. If $t : T \to T'$ is a natural transformation into another functor T', not necessarily proper, then

$$t(A)(x) = t(A)\big(T(f)(y)\big) = T'(f)\big(t(D)(y)\big).$$

Thus t is completely determined by its restriction $t/_{\underline{D}}$. This means of course that $t/_D = t'/_D$ implies $t = t'$.

By $[T, T']$, we shall mean the set

$$[T/_{\underline{D}}, T'] \subset \Pi_{D \text{ in } \underline{D}} [T(D), T'(D)].$$

In this way, the class of all proper functors from $\underline{A}$ to Ens is turned into a category that may again be denoted by $[\underline{A}, \text{Ens}]$. Actually, this category depends, in a technical sense, on the choice of a dominating set for each small functor. If $\underline{A}$ is small, we choose all dominating sets to be $\underline{A}$, so that $[\underline{A}, \text{Ens}]$ then has the same meaning as before.

We shall also write $[\underline{A}, \text{Ens}]_{\text{inf}}$ for the category of all proper inf-preserving functors from $\underline{A}$ to Ens. We observe that, for any object A of $\underline{A}$, the functor $[-, A] : \underline{A}^{\circ} \to \text{Ens}$ is proper, with dominating set $\{A\}$. Therefore we still have canonical embeddings $\underline{A} \to [\underline{A}^{\circ}, \text{Ens}]$ and $\underline{A} \to [\underline{A}^{\circ}, \text{Ens}]_{\text{inf}}$.

The reader will recall the category ${X_T}^{\circ}$ and the diagram ${\Gamma_T}^{\circ} : {X_T}^{\circ} \to \underline{A}$ associated with any functor $T : \underline{A}^{\circ} \to \text{Ens}$ from Section 2. If $T = [G -, B]$, where $G : \underline{A} \to \underline{B}$ and B is an object of $\underline{B}$, they were discussed in the proof of Proposition 2.1. We now state a generalization of the corollary to the latter.

PROPOSITION 5.1. Let there be given a functor $G : \underline{A} \to \underline{B}$ such that $[G -, B] : \underline{A}^{\circ} \to \text{Ens}$ is proper for each object B of $\underline{B}$. Let X_B and $\Gamma_B : X \to \underline{A}$ be the category and diagram associated with $[G -, B]$. Then the functor $B \rightsquigarrow [G -, B] : \underline{B} \to [\underline{A}^{\circ}, \text{Ens}]$ is an embedding if and only if, for each B in $\underline{B}$, sup $(G \circ \Gamma_B) = (B, 1)$, where $1(x) = x$ for each x in X_B.

We may call $G : \underline{A} \to \underline{B}$ *left adequate* if it satisfies the equivalent conditions of this proposition. This definition agrees with the earlier one when $\underline{A}$ is small and corresponds to Isbell's term "properly left adequate". *Right adequate* functors are defined dually.

Proof. First, assume that $B \rightsquigarrow [G -, B]$ is faithful and full, we wish to show that sup $(G \circ \Gamma_B) = (B, 1)$. Let $t(x) : G(\Gamma_B(x)) \to B'$ naturally in x, we seek a unique $b : B \to B'$ such that $bx = t(x)$.

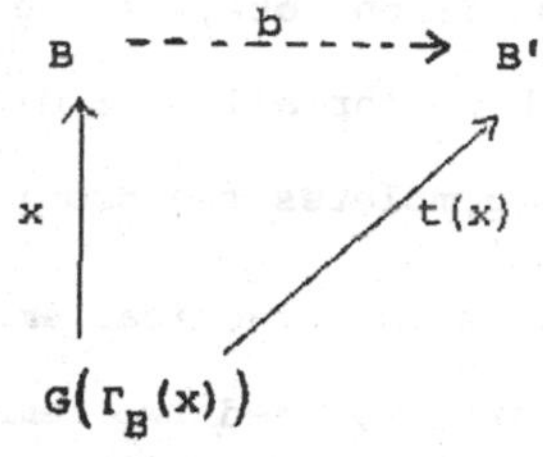

Define $t'(A) : [G(A), B] \to [G(A), B']$ by $t'(A)(x) = t(x)$. One easily verifies that $t'(A)$ is natural in A. Write $t' = \mu(t)$, then $\mu : [G \circ \Gamma_B, B'_{X_B}] \to \big[[G -, B], [G -, B']\big]$. By assumption, the mapping $b \rightsquigarrow [G -, b] : [B, B'] \to \big[[G -, B], [G -, B']\big]$ has an inverse, call it λ. Put $b = \lambda(\mu(t))$, i.e., $[G -, b] = t'$, that is to say

$$bx = [G(A), b](x) = t'(A)(x) = t(x).$$

b is unique with this property; for if also $b'x = t(x)$, then $[G -, b'] = t'$, hence $b' = \lambda(t') = b$. This completes the first part of the proof.

Conversely, assume that $\sup(G \circ \Gamma_B) = (B, 1)$.
We will show that the mapping $[B, B'] \to \big[[G -, B], [G -, B']\big]$ is one-one and onto.

Let $t' : [G -, B] \to [G -, B']$. Then, for any $x \in [G(\Gamma_B(x)), B]$, we have $t'(\Gamma_B(x)) : G(\Gamma_B(x)) \to B'$. One easily verifies that this is natural in x. By assumption, there exists a unique $b : B \to B'$ such that

$$t'(\Gamma_B(x))(x) = bx = \big[G(\Gamma_B(x)), b\big](x).$$

Thus $t' = [G -, b]$, and so our mapping is onto.

To see that the mapping is one-one, assume that also $t' = [G -, b']$. Then $bx = b'x$ for all $x : G(A) \to B$, hence $b' = b$ by uniqueness. This completes the proof.

Before we can assert that the canonical embedding $\underline{A} \to [\underline{A}^o, \text{Ens}]$ is left adequate, we need two lemmas.

LEMMA 5.1. Let T and T' be isomorphic functors from $\underline{A}^o$ to Ens. If T is proper then so is T', with the same dominating set.

The proof is routine and will be omitted.

LEMMA 5.2. Any functor $F : \underline{C} \to \underline{C}$ which is isomorphic to the identity functor of $\underline{C}$ is an embedding.

Proof. Let $t(C) : C \to F(C)$ be the given isomorphism, natural in C. Then, for any map $c : C \to C'$, $F(c) = t(C')c\, t(C)^{-1}$. It follows that the mapping $c \rightsquigarrow F(c) : [C, C'] \to [F(C), F(C')]$ is one-one and onto.

PROPOSITION 5.2. For any category $\underline{A}$, the canonical embeddings $\underline{A} \to [\underline{A}^o, \text{Ens}]$ and $\underline{A} \to [\underline{A}^o, \text{Ens}]_{inf}$ are left adequate, sup-dense and inf-preserving, the second is also sup-preserving. Moreover $[\underline{A}^o, \text{Ens}]$ is sup-complete.

The last statement has also been asserted by Benabou(1965).

__Proof.__ Consider the embedding $H^o : \underline{A} \to [\underline{A}^o, \text{Ens}]$. Take any T in $[\underline{A}^o, \text{Ens}]$, then $[H^o-, T] \cong T$, by Yoneda's Lemma. By Lemma 5.1, $[H^o-, T]$ is proper. By Proposition 5.1 H^o is left adequate and sup-dense. By the corollary to Proposition 3.2, H^o preserves infs. The statements concerning the other embedding are proved similarly.

It remains to show that $[\underline{A}^o, \text{Ens}]$ is sup-complete. Let $\Gamma : I \to [\underline{A}^o, \text{Ens}]$ be a diagram with small index category I. For each i in I, $\Gamma(i) : \underline{A}^o \to \text{Ens}$ is proper, let us say with dominating set $\underline{D}_i$. For any A in $\underline{A}$, write $\Gamma'(A)(i) = \Gamma(i)(A)$ and verify that $\Gamma'(A) : I \to \text{Ens}$ is a functor. (See the proof of Proposition 1C.)

Let $\sup \Gamma'(A) = \big(T(A), u'(A)\big)$, and write $u(i)(A) = u'(A)(i)$. As in the proof of Proposition 1C, T is a functor and u is a natural transformation. Moreover, it will follow that $\sup \Gamma = (T, u)$, if we make sure that T is in $[\underline{A}^o, \text{Ens}]$ at all. We claim that T is proper with dominating set $\underline{D} = \cup_{i \in I} \underline{D}_i$.

Indeed, let $x \in T(A)$, then $x \in v(i)(A)\big(\Gamma(i)(A)\big)$, for some i in I. (Otherwise there would exist a mapping $T(A) \to T(A) - \{x\}$ such that $gv(i)(A) = v(i)(A)$, leading to a contradiction.) Hence $x = v(i)(A)(z)$, where $z \in \Gamma(i)(A)$. Therefore there exist D in $\underline{D}_i$, the dominating set of $\Gamma(i)$, $y \in \Gamma(i)(D)$, and $f : A \to D$ such that $z = \Gamma(i)(f)(y)$, hence

$$x = v(i)(A)(z) = \big(v(i)(A)\Gamma(i)(f)\big)(y) = \big(T(f)v(i)(D)\big)(y),$$

by naturality, and so $x = T(f)(y')$, where $y' = v(i)(D)(y) \in T(D)$. This completes the proof.

Unfortunately, we could not show that $[\underline{A}^{\circ}, \text{Ens}]$ and $[\underline{A}^{\circ}, \text{Ens}]_{\text{inf}}$ are inf-complete. Instead, we shall establish a kind of representation completeness. But first we require a somewhat technical result. (The small circle may be used to denote the product of functors.)

LEMMA 5.3. Let $G : \underline{A} \to \underline{B}$ be left adequate and $F, F' : \underline{B} \to \underline{C}$ with F sup-preserving. Then every natural transformation $t : F \circ G \to F' \circ G$ can be extended to a unique natural transformation $t' : F \to F'$ so that $t'\big(G(A)\big) = t(A)$. If also F' preserves sups, then $F \circ G \cong F' \circ G$ implies $F \cong F'$.

Proof. We recall from Proposition 5.1 that $\sup\,(G \circ \Gamma_B) = (B, 1)$, hence $\sup\,(F \circ G \circ \Gamma_B) = \big(F(B), F \circ 1\big)$. Since $F'(x)t\big(\Gamma_B(x)\big) : F\big(G(\Gamma_B(x))\big) \to F'(B)$, there exists a unique $t'(B) : F(B) \to F'(B)$ such that $t'(B)F(x) = F'(x)t\big(\Gamma_B(x)\big)$. It is easily shown that $t'(B)$ is natural in B. Take x to be the identity map of $G(A)$, then $B = G(A)$ and $\Gamma_B(x) = A$, hence $t'\big(G(A)\big) = t(A)$.

Next, suppose that both F and F' preserve sups and that $t : F \circ G \to F' \circ G$ is a natural isomorphism with inverse u.

Extend t to $t' : F \to F'$ and u to $u' : F' \to F$. Then $u't' : F \to F$ and

$$(u't')(G(A)) = u'(G(A))t'(G(A)) = u(A)t(A) = 1$$

Thus $u't'$ extends $1 : F \circ G \to F \circ G$. But so does $1 : F \to F$, hence $u't' = 1$. Similarly $t'u' = 1$, and our proof is complete.

PROPOSITION 5.3. Every proper inf-preserving functor from $[\underline{A}, \text{Ens}]^{o}$ or $[\underline{A}, \text{Ens}]_{\text{inf}}{}^{o}$ to Ens is representable.

Proof. Take for example the latter category. Let $F : [\underline{A}, \text{Ens}]_{\text{inf}}{}^{o} \to \text{Ens}$ be inf-preserving and proper. As before, consider the canonical embedding $H : \underline{A} \to [\underline{A}, \text{Ens}]_{\text{inf}}{}^{o}$ defined by $H(A) = [\,-, A]$. Then H preserves infs (see Proposition 5.2), hence so does $F \circ H : \underline{A} \to \text{Ens}$. Let us assume for the moment that $F \circ H$ is proper, then it follows that $F \circ H$ is an object of $[\underline{A}, \text{Ens}]_{\text{inf}}{}^{o}$. Therefore $F \circ H \cong [F \circ H, H\,-]$, by Yoneda's Lemma. Putting $F' = [F \circ H, -]$, we thus have $F \circ H \cong F' \circ H$. Now H is right adequate (see Proposition 5.2) and both F and F' preserve infs. We may therefore apply Lemma 5.3 and deduce that $F \cong F' = [F \circ H, -]$. Thus F is representable.

It remains to show that $F \circ H$ is proper. Let $\underline{D}$ be a dominating set for F. Then $\underline{D}$ is a subcategory of $[\underline{A}, \text{Ens}]_{\text{inf}}{}^{o}$, and any object of $\underline{D}$ is itself a proper functor $D : \underline{A} \to \text{Ens}$, let us say with dominating set $\underline{E}_D$. We claim that $\underline{E} = \cup_{D \text{ in } \underline{D}} \underline{E}_D$ is a dominating set for $F \circ H$.

Indeed, take any A in $\underline{A}$ and $x \in F(H(A))$. Since F has dominating set $\underline{D}$, there exist D in $\underline{D}$, $z \in F(D)$, and $g : D \to H(A)$ such that $x = F(g)(z)$.

Now $g \in [D, H(A)]$ corresponds to some $g' \in D(A)$ under the Yoneda isomorphism $[D, H(A)] \cong D(A)$. Since D has dominating set $\underline{E}_D$, we may pick E in $\underline{E}_D$, $h' \in D(E)$, and $f : E \to A$ such that $g' = D(f)(h')$. Again, h' corresponds to some $h \in [D, H(E)] \cong D(E)$. The relation between g and h is clearly this:

$$g = [D, H(f)](h) = H(f)h,$$

as illustrated by the commutative diagram:

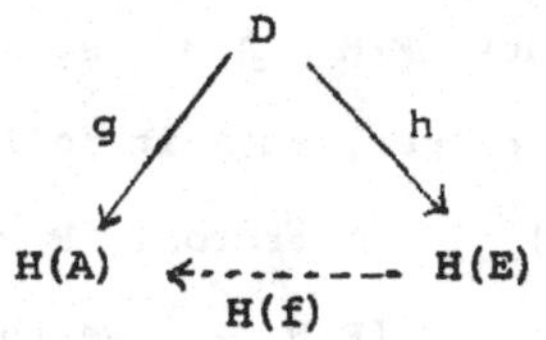

Thus

$$x = F(g)(z) = \big(F(H(f))F(h)\big)(z) = F\big(H(f)\big)(y),$$

where $y = F(h)(z) \in F\big(H(E)\big)$, and E is in $\underline{E}$, as required.

6. The relationship between different forms of completeness. We aim to investigate the relationship between inf- and sup-completeness on the one hand and the representability of functors on the other. The following result has also been announced by Benabou (1965).

PROPOSITION 6.1. If $\underline{A}$ is inf-complete then every proper inf-preserving functor $\underline{A} \to \text{Ens}$ is representable.

Proof. Recall that the canonical embedding $H : \underline{A} \to [\underline{A}, \text{Ens}]_{\text{inf}}^{\circ}$ is right adequate. With any functor $T : \underline{A} \to \text{Ens}$ we associate the category $X = X_T$ and the diagram $\Gamma = \Gamma_T : X \to \underline{A}$ as in Section 2. By Proposition 5.1, or rather its dual, we have $\inf (H \circ \Gamma) = (T, 1)$. (Actually, the quoted proposition deals not with T, but with the isomorphic functor $[T, H -] : \underline{A} \to \text{Ens}$, where the Hom is taken in $[\underline{A}, \text{Ens}]_{\text{inf}}$.)

Now $\underline{A}$ is inf-complete and H preserves infs. We might try to argue that $\inf \Gamma = (A, u)$ in $\underline{A}$, hence $T \cong H(A)$. The only trouble with this argument is that X is not small. However, let $\underline{D}$ be a dominating set for T, and let Y be the subcategory of X whose objects form the set $\cup_{D \text{ in } \underline{D}} T(D)$. Then Y is a small category and $\inf \Gamma/_Y = (A, u)$, say. As we shall see, this implies $\inf \Gamma = (A, v)$, for suitable v, hence $\inf (H \circ \Gamma) = \big(H(A), H \circ v\big)$, and so $T \cong H(A)$, as required.

It remains to show that $\inf \Gamma = (A, v)$. To this purpose it suffices to extend the natural transformation $u : \Gamma/_Y \to A_Y$

to $v : \Gamma \to A_X$. Let us ask more generally:

PROBLEM. Suppose Y is a subcategory of some category X, $\Gamma : X \to \underline{A}$, $u : \Gamma/_Y \to A_Y$ a lower bound of $\Gamma/_Y$. When can u be extended to a lower bound $v : \Gamma \to A_X$?

The following pair of conditions is sufficient:

CONDITION I. For all x in X there exist y in Y and a map $\xi : y \to x$.

CONDITION II. If $\xi_1 : y_1 \to x$ and $\xi_2 : y_2 \to x$, then there exist x' in X, $\mu_1 : x' \to y_1$, $\mu_2 : x' \to y_2$ such that $\xi_1\mu_1 = \xi_2\mu_2$.

The first condition allows us to define $v(x) = \Gamma(\xi)\, u(y)$, and the second condition assures that the definition does not depend on the choice of ξ and y.

We now continue with the proof of Proposition 6.1. Condition I is satisfied since T is proper. To verify Condition II, assume that $\xi_i : y_i \to x$, $i = 1$ or 2, where $\xi_i = (y_i, x, a_i)$, $y_i \in T(D_i)$, $x \in T(A)$, $a_i : D_i \to A$, and $x = T(a_i)(y_i)$. We seek x' and $\mu_i : x' \to y_i$ such that $\xi_1\mu_1 = \xi_2\mu_2$.

Since $\underline{A}$ is inf-complete and T preserves infs, we may form the pullbacks below. (For the moment, disregard R and the arrows emanating from it.)

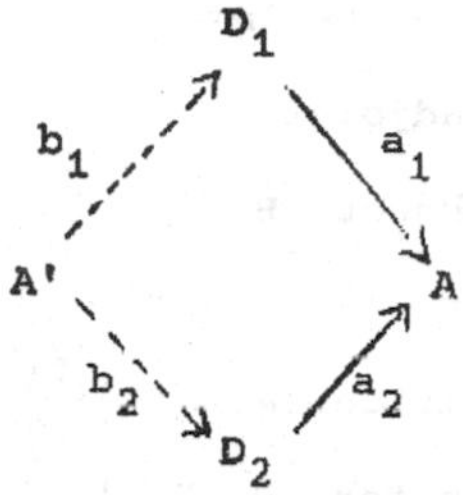

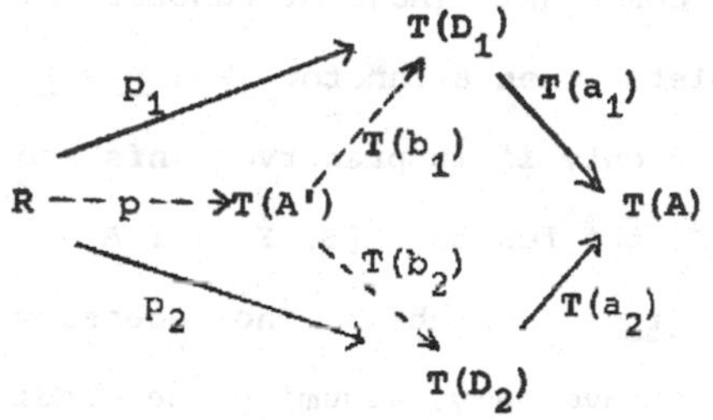

Now let $R \subset T(D_1) \times T(D_2)$ be defined by

$$R = \{(z_1, z_2) \mid T(a_1)(z_1) = T(a_2)(z_2)\}.$$

Also let $p_i : R \to T(D_i)$ be given by $p_i\big((z_1, z_2)\big) = z_i$. Then there exists a unique $p : R \to T(A')$ such that $T(b_i)p = p_i$. Since $(y_1, y_2) \in R$, we may put $x' = p\big((y_1, y_2)\big)$, hence

$$y_i = p_i\big((y_1, y_2)\big) = \big(T(b_i)p\big)\big((y_1, y_2)\big) = T(b_i)(x').$$

Putting $\mu_i = (x', y_i, b_i)$, we then have

$$\xi_1\mu_1 = (x', x, a_1b_1) = (x', x, a_2b_2) = \xi_2\mu_2,$$

as required.

This completes the proof of Proposition 6.1. Of course one can also prove this directly, without recourse to Proposition 5.1, by taking a suitable inf. Such a procedure has in fact been proposed by Lawvere for the general Adjoint Functor Theorem, which will here be obtained as a corollary, as is also done by Benabou.

COROLLARY. (Adjoint Functor Theorem.) Let $\underline{A}$ be inf-complete, then a functor $F : \underline{A} \to \underline{B}$ has a left adjoint if and only if it preserves infs and, for each object B of $\underline{B}$, the functor $[B, F -] : \underline{A} \to \text{Ens}$ is proper.

<u>Proof</u> (sketched). The necessity of the condition is easy. Conversely, assuming the condition, the functor $[B, F -]$ may be represented by some object $G(B)$ of $\underline{A}$, by Proposition 6.1. It readily follows that G is a functor and a left adjoint of F.

Proposition 6.1 admits a pseudo-converse, in which inf-completeness is replaced by a form of sup-completeness. Let us call a diagram $\Gamma : I \to \underline{A}$ <u>proper</u> when the associated functor $A \rightsquigarrow [\Gamma, A_I]$ is proper. This means that $\underline{A}$ has a small subcategory $\underline{D}$ so that, for every upper bound $x : \Gamma \to A_I$, there exists an upper bound $y : \Gamma \to D_I$, D in $\underline{D}$, and a map $f : D \to A$ such that $x(i) = f\, y(i)$, for all i in I.

PROPOSITION 6.2. Suppose every proper inf-preserving functor $\underline{A} \to \text{Ens}$ is representable. Then a diagram $\Gamma : I \to \underline{A}$ has a sup if and only if it is proper.

<u>Proof.</u> If $\sup \Gamma = (D, y)$, the functor $A \rightsquigarrow [\Gamma, A_I]$ is indeed proper, with dominating set $\{D\}$.

Conversely, assume that the functor $A \rightsquigarrow [\Gamma, A_I]$ is proper. Now it is easily seen that it preserves infs. Therefore it is representable, by Proposition 6.1. Thus there exists an object B of $\underline{A}$ such that $[\Gamma, A_I] \cong [B, A]$, naturally in A. By Proposition 1A, this means that $\sup \Gamma = B$.

COROLLARY. An inf-complete category is sup-complete if and only if every diagram $I \to \underline{A}$ with small index category I is proper.

Propositions 6.1 and 6.2 together almost establish the equivalence of inf-completeness and sup-completeness, but not quite. To rescue something from this situation we shall make a definition. Recall that every proper functor $T : \underline{A} \to Ens$ gives rise to a functor $T^* : \underline{A}^o \to Ens$, where $T^*(A) = [H(A), T]$ in $[\underline{A}, Ens]^o$. Dually, every proper functor $U : \underline{A}^o \to Ens$ gives rise to a functor $U^+ : \underline{A} \to Ens$, where $U^+(A) = [U, H^o(A)]$ in $[\underline{A}^o, Ens]$. We shall call the functor T <u>very proper</u> if the functors

$$T, T^*, T^{*+}, T^{*+*}, \ .....$$

are all proper. (Each of them exists because the preceding one is proper.) We call the diagram $\Gamma : I \to \underline{A}$ <u>very proper</u> if the index category I is small and the associated functor $A \rightsquigarrow [\Gamma, A_I]$ is very proper.

PROPOSITION 6.3. Given any category $\underline{A}$, the following conditions on $\underline{A}$ are equivalent:

(1) Every very proper diagram $I \to \underline{A}$ has an inf.

(2) Every very proper inf-preserving functor $\underline{A} \to Ens$ is representable.

(3) Every very proper diagram $I \to \underline{A}$ has a sup.

(4) Every very proper inf-preserving functor $\underline{A}^{o} \to \mathrm{Ens}$ is representable.

<u>Proof.</u> In view of duality considerations, it will suffice to show (1) → (2) → (3). We shall temporarily write $\underline{T}$ for the opposite category of all very proper inf-preserving functors from $\underline{A}$ to Ens.

Assume (1) and let T be in $\underline{T}$. Form the associated category $X = X_T$ and diagram $\Gamma = \Gamma_T : X \to \underline{A}$, as in Section 2. Now the functor $H : \underline{A} \to \underline{T}$ is still right adequate (see Lemma 6.1 below). Moreover, $[T, H-] \cong T$ in $[\underline{A}, \mathrm{Ens}]^{o}$ is proper for each T in $\underline{T}$, by Lemma 5.1. Therefore, by the dual of Proposition 5.1, $\inf (H \circ \Gamma) = T$. As in the proof of Proposition 6.1, X has a small subcategory Y such that $\inf \Gamma = \inf \Gamma/_{Y}$, if the latter exists. That it does exist will follow from (1) if we show that $\Gamma/_{Y}$ is very proper. Assuming this for the moment, put $\inf \Gamma = B$. Since H preserves infs, $\inf (H \circ \Gamma) = H(B)$. Therefore $T \cong H(B)$, and so (2).

We must still show that $\Gamma/_{Y}$ is very proper. Let $U : \underline{A}^{o} \to \mathrm{Ens}$ be the functor defined by $U(A) = [A_Y, \Gamma/_{Y}]$. Consider any $t \in U(A)$, then $t(y) \in [A, \Gamma(y)] \cong \big[H(A), H\big(\Gamma(y)\big)\big]$ in $\underline{T}$. Under this isomorphism, $t(y)$ goes to $H\big(t(y)\big) = (H \bullet t)(y)$. It is easily verified that $t \rightsquigarrow H \bullet t$ is an isomorphism of $[A_Y, \Gamma/_{Y}]$ to $[H(A)_Y, (H \circ \Gamma)/_{Y}]$. But the latter is isomorphic to $[H(A), T]$, by Proposition 1A. One easily verifies naturality in A, hence $U \cong [H -, T] \cong T^*$. Since T is very proper, so is T^*, according to the definition of "very proper", hence also U, by Lemma 6.2 below. This completes the proof that (1) $\rightarrow$ (2).

Assume (2), and let $\Gamma : I \rightarrow \underline{A}$ be a very proper diagram. This means that $T : \underline{A} \rightarrow$ Ens is very proper, where $T(A) = [\Gamma, A_I]$. Since T evidently preserves infs, there is an object B in $\underline{A}$ such that $T \cong [B, -]$. Therefore $[\Gamma, A_I] \cong [B, A]$, naturally in A, hence $\sup \Gamma = B$, by Proposition 1A. Thus (2) $\rightarrow$ (3), and our proof is complete, subject to the two lemmas below.

LEMMA 6.1. Given a functor $J : \underline{A} \rightarrow \underline{B}$, where $\underline{B}$ is a subcategory of $\underline{C}$. If the composite functor $\underline{A} \rightarrow \underline{B} \rightarrow \underline{C}$ is left adequate, then so is J.

The proof is trivial and will be omitted.

LEMMA 6.2. If $S \cong T : \underline{A} \rightarrow$ Ens, then $S^* \cong T^* : \underline{A}^o \rightarrow$ Ens, and S is very proper if and only if T is very proper.

Proof. Let $u : S \rightarrow T$, then $u^* : S^* \rightarrow T^*$, where $u^* = [H -, u]$. It is easily seen that u^* is a natural transformation. Moreover, if $v : T \rightarrow S$ is the inverse of u, then $u^*v^* = (uv)^* = 1^* = 1$, and similarly $v^*u^* = 1$. Thus $S^* \cong T^*$. In this way, also $S^{*+} \cong T^{*+}$, $S^{*+*} \cong T^{*+*}$, etc.

Now assume that S is very proper. This means that $S, S^*, S^{*+}, \ldots$ are all proper. By Lemma 5.1, also $T, T^*, T^{*+}, \ldots$ are all proper. Thus T is very proper.

7. Theorems without properness conditions.

We wish to investigate when the properness conditions on Proposition 6.1 and the corollary to Proposition 6.2 can be removed. The proof of the following has been adapted from Mitchell's proof of the Special Adjoint Functor Theorem [Mitchell, page 126].

PROPOSITION 7.1. Let $\underline{A}$ be inf-complete. Then every inf-preserving functor $T : \underline{A} \to \mathrm{Ens}$ is proper, hence representable, in either of the following two cases:

CASE 1. $\underline{A}$ contains a right adequate small subcategory $\underline{C}$.

CASE 2. $\underline{A}$ contains a cogenerating small subcategory $\underline{C}$, and every object of $\underline{A}$ has a representative set of subobjects.

Under these assumptions, every inf-preserving functor $F : \underline{A} \to \underline{B}$ has a left adjoint.

Proof. Given any object A of $\underline{A}$, let $I = X_{[A,\,-]}$ and $\Delta = \Gamma_{[A,\,-]} : X \to \underline{C} \to \underline{A}$ be the category and diagram associated with the functor $[A, -] : \underline{C} \to \mathrm{Ens}$, as in Section 2. (Actually, we should put $\Delta : X \to \underline{C}$, but we may as well let it absorb the inclusion functor.) In Case 1, $\inf \Delta = (A, 1)$, where $1(i) = i$ for all i in I, by the dual of Proposition 5.1. In Case 2, we apply the dual of Proposition 2.1, or rather its proof, and obtain $\inf \Delta = (A', w)$, with a monomorphism $k : A \to A'$ such that $w(i)k = i$ for all i in I.

In either case, let $Y = X_{T/\underline{C}}$ and $\Gamma = \Gamma_{T/\underline{C}} : I \to \underline{C} \to \underline{A}$ be the category and diagram associated with the functor $T/_{\underline{C}}$. (As above, the inclusion functor has been absorbed in Γ.) Put $\inf \Gamma = (B, u)$, where B is in $\underline{A}$.

Take any $x \in T(A)$. For any $i : A \to C = \Delta(i)$, put $y_i = T(i)(x) \in T(C)$. Write $v_x(i) = u(y_i) : B \to \Delta(i)$ and verify that this is natural in i. In Case 1, there exists a unique $a : B \to A$ such that $ia = v_x(i) = u(y_i)$. In Case 2, there exists a unique $a : B \to A'$ such that $w(i)a = u(y_i)$.

Since T preserves infs, $\inf (T \circ \Gamma) = \big(T(B), T \circ u\big)$, hence there exists a unique $\hat{z} : \{O\} \to T(B)$ such that $T\big(u(y)\big)\, \hat{z} = \hat{y}$, for all y in Y. (Here $\hat{y}\,(O) = y$, see the Remark in Section 2.)

In Case 1, there exists a unique map $f : \{O\} \to T(A)$ such that $T(i) f = \hat{y}_i$. Now on the one hand $\hat{y}_i = T(i)\hat{x}$, and on the other hand $\hat{y}_i = T\big(u(y_i)\big)\hat{z} = T(i)T(a)\hat{z}$. Therefore $\hat{x} = f = T(a)\hat{z}$, and so $x = T(a)(z)$, $z \in T(B)$. Thus $\{B\}$ is a dominating set for T.

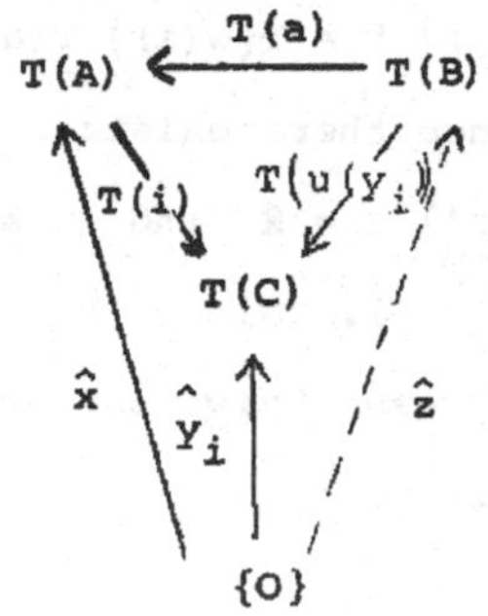

In Case 2, form the pullback:

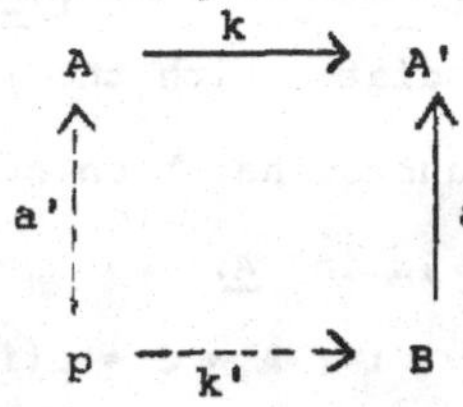

Then k' is easily seen to be a monomorphism. Since T preserves infs, the square in the following diagram is another pullback:

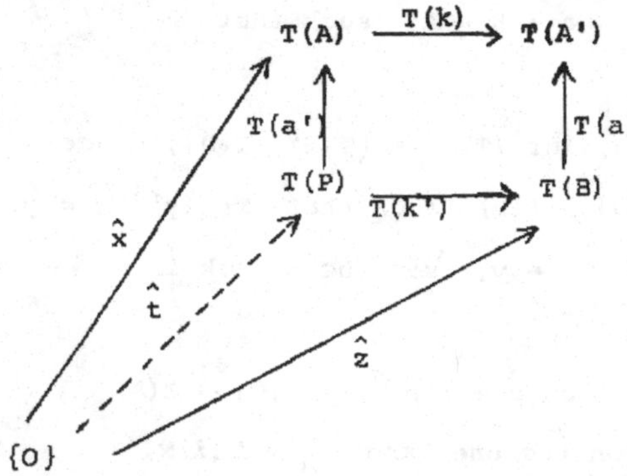

Again, there exists a unique $f : \{0\} \to T(A')$ such that $T(w(i))f = \hat{y}_i$. On the one hand $\hat{y}_i = T(i)\,\hat{x} = T(w(i))T(k)\hat{x}$. On the other hand $\hat{y}_i = T(u(y_i))\,\hat{z} = T(w(i))\,T(a)\hat{z}$. Therefore $T(k)\hat{x} = T(a)\hat{z}$. Hence there exists a unique $\hat{t} : \{0\} \to T(P)$ such that $T(a')\,\hat{t} = \hat{x}$ and $T(k')\,\hat{t} = \hat{z}$. Thus $T(a')(t) = x$ with $t \in T(P)$. Since P is a subobject of B, it follows that any representative set of subobjects of B is a dominating set for T.

Before stating our next result, we summarize some ideas of Isbell (1964). He called a monomorphism m extremal if $m = m'e'$ and e' epi implies that e' is an isomorphism. He proved this:

PROPOSITION 7A. If $\underline{A}$ is inf-complete and every object of $\underline{A}$ has a representative set of subobjects, then every map f of $\underline{A}$ has a canonical decomposition $f = f_m f_e$, where f_e is epi and f_m is an extremal monomorphism, and this decomposition is unique up to isomorphism. Moreover, the product of extremal monomorphisms is again an extremal monomorphism.

From this we deduce the following:

LEMMA 7.1. (Diagonal Lemma.) Assume that $\underline{A}$ is inf-complete and every object has a representative set of subobjects. If $mg = he$, m an extremal monomorphism and e epi, then there exists a unique d such that $md = h$.

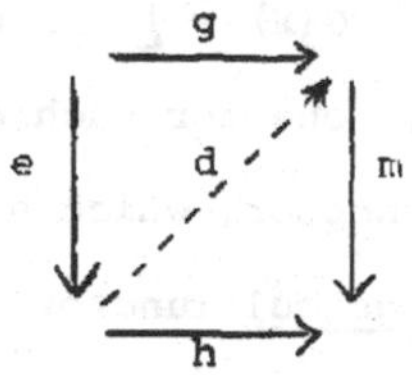

Proof. Let $g = g_m g_e$ and $h = h_m h_e$ be the canonical decompositions. Then $(mg_m)g_e = h_m(h_e e) = f$, say, are two canonical decompositions of f. By uniqueness, there exists an isomorphism x such that $xg_e = h_e e$ and $h_m x = mg_m$.

Take $d = g_m x^{-1} h_e$, then $md = h$. Since m is mono, d is unique with this property.

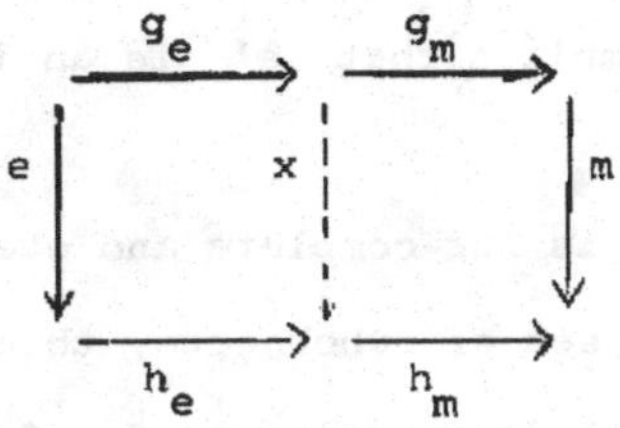

PROPOSITION 7.2. Let $\underline{A}$ be inf-complete. Then $\underline{A}$ is also sup-complete in either of the following two cases:

CASE 1. $\underline{A}$ contains a left adequate subcategory consisting of a single object A_o, and arbitrary sums of A_o exist in $\underline{A}$.

CASE 2. $\underline{A}$ contains a generator A_o, and arbitrary sums of A_o exist in $\underline{A}$. Moreover, every object in $\underline{A}$ has representative sets of subobjects and quotient objects.

Proof. For any set X, let $G(X) = \sum_{x \in X} A_o$ denote the direct sum of copies of A_o, one for each element $x \in X$. Then $G : Ens \to \underline{A}$ is a functor, which has as a right adjoint the so-called forgetful functor $F = [A_o, -] : \underline{A} \to Ens$. Indeed,

$$[X, F(A)] \cong \Pi_{x \in X}[\{x\}, F(A)] \cong \Pi_{x \in X} [A_o, A] \cong [\textstyle\sum_{x \in X} A_o, A] = [G(X), A].$$

It follows that there exist canonical natural transformations $e : G \circ F \to 1$ and $m : 1 \to F \circ G$ with well-known universal properties.

Consider any diagram $\Gamma : I \to \underline{A}$, where I is small. In view of the corollary to Proposition 6.2, we need only show that Γ is proper. Thus we want to find a small subcategory $\underline{D}$ of $\underline{A}$ so that, given $x : \Gamma \to A_I$, there exist D in $\underline{D}$, $y : \Gamma \to D_I$, and $n : D \to A$ such that $x(i) = n\, y(i)$ for all i in I.

Let $\sup\,(F \circ \Gamma) = (V, v)$. Now G preserves sups (see the corollary to Proposition 3A), hence also $\sup\,(G \circ F \circ \Gamma) = \big(G(V), G(v)\big)$. Since $x(i) e\big(\Gamma(i)\big) : G\Big(F\big(\Gamma(i)\big)\Big) \to A$ is easily seen to be natural in i, there exists a unique map $f : G(V) \to A$ such that $f\, G\big(v(i)\big) = x(i)\; e\big(\Gamma(i)\big)$.

We shall use the following facts:

(a) $e(A) : G\big(F(A)\big) \to A$ is epi.

(b) $F\big(e(A)\big) : F\Big(G\big(F(A)\big)\Big) \to F(A)$ is epi.

(a) is an easy consequence of the fact that A_0 is a generator. (b) is not deduced from (a), but is shown directly. Indeed, let $x \in F\big(G(A)\big) = [A_0, A] = \Big[G\big(\{0\}\big), A\Big]$. Then, by the universal property of e, there exists a unique $f : \{0\} \to F(A)$ such that

$$x = e(A) G(f) = [A_0, e(A)]\big(G(f)\big) = F\big(e(A)\big)\big(G(f)\big).$$

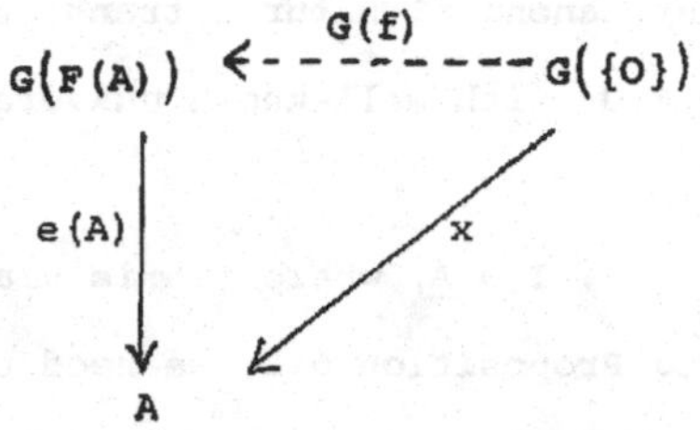

In view of the preliminary spadework done by Isbell, Case 2 will be a little easier to deal with than Case 1. We shall therefore consider it first. Let $f = me$ with image D, where m is an extremal monomorphism and e is epi. Since $e(\Gamma(i))$ is epi (see (a) above), we may apply the Diagonal Lemma (Lemma 7.1), and obtain a unique map $y(i) : \Gamma(i) \to D$ such that $m\, y(i) = x(i)$. It is easily verified that $y(i)$ is natural in i, hence $y : \Gamma \to D_I$. Thus we may take $\underline{D}$ to be a representative set of quotient objects of $G(V)$.

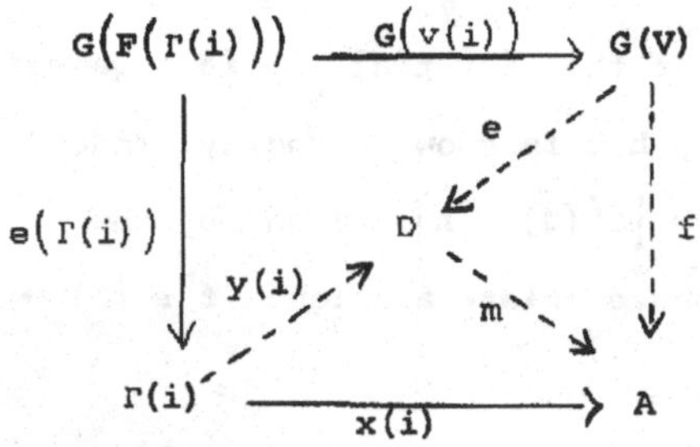

In Case 1 we shall consider the same square as above, but after applying the forgetful functor F.

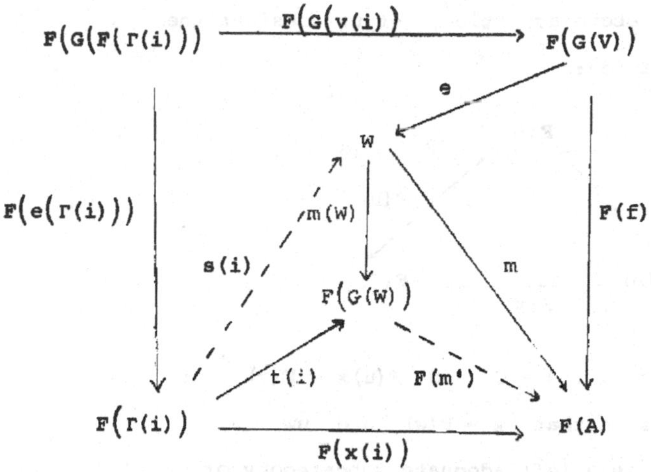

Write $F(f) = me$, where m is mono and e is epi in Ens, with image W. Since $F(e(\Gamma(i)))$ is epi (see (b) above), there exists a unique mapping $s(i)$ such that $m\,s(i) = F(x(i))$. It may be verified that $s(i)$ is natural in i.

Now consider the canonical map $m(W) : W \to F(G(W))$. By the universal property of m, there exists a unique $m' : G(W) \to A$ such that $F(m')m(W) = m$. Write $t(i) = m(W)\ s(i)$, then $F(m')t(i) = m\ s(i) = F(x(i))$. Our argument will be complete if we can assert that $t(i) = F(y(i))$ such that $m'y(i) = x(i)$. For we may then take the dominating set $\underline{D}$ to be the set of all $G(W)$, where W is any quotient set of $F(G(V))$, let us say, defined by an equivalence relation.

To prove the above assertion, let us consider the more general situation:

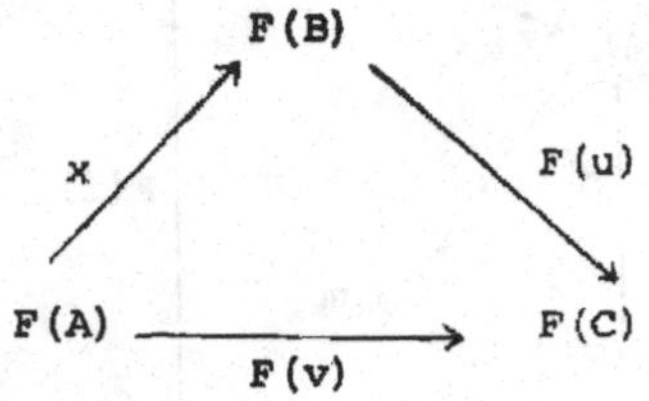

It is assumed that $u : B \to C$ and $F(u)x = F(v)$. We seek $w : A \to B$ such that $x = F(w)$ and $uw = v$.

Since $\{A_0\}$ is a left adequate subcategory of $\underline{A}$, the mapping $F : [A, B] \to [F(A), F(B)]$ is one-one and onto. Because it is onto, there exists $w : A \to B$ such that $x = F(w)$. Therefore $F(uw) = F(u)x = F(v)$. Because F is one-one, $uw = v$. This completes the proof.

For case 2, see also Benabou (1965), Théorème 5.

8. Completions of groups. We wish to investigate $[\underline{A}^{o}, \mathrm{Ens}]_{inf}$ when $\underline{A}$ is some known small category. In our first example we take $\underline{A}$ to be a group G : $\underline{A}$ has one object, we may as well call it G, and the maps of $\underline{A}$ (= elements of G) are all isomorphisms.

Now consider any small index category I and a diagram $\Gamma : I \to \underline{A}$. We call Γ and I connected if, given any two objects $i, j \in I$, there exist objects

(*) $\quad i_o = i, i_1, i_2, \ldots, i_n = j$

such that, for any $k = 0, 1, . . ., n - 1$, one of $[i_k, i_{k+1}]$ or $[i_{k+1}, i_k]$ is nonempty.

What do the lower bounds of a connected diagram Γ look like? Suppose (G, s) is a lower bound of Γ. Consider two neighbouring indices i_k, i_{k+1}, so that $[i_k, i_{k+1}]$ or $[i_{k+1}, i_k]$ is not empty. In the first case we have a map $\iota_k : i_k \to i_{k+1}$. By naturality, $\Gamma(\iota_k)\, s(i_{k+1}) = s(i_k)$, hence $s(i_{k+1}) = \Gamma(\iota_k)^{-1}\, s(i_k)$. In the second case we may take $\iota_k : i_{k+1} \to i_k$ and $s(i_{k+1}) = \Gamma(\iota_k) s(i_k)$. Thus in either case $s(i_{k+1})$ is determined by $s(i_k)$. Applying this to the sequence (*), we obtain

(**) $\quad s(j) = \Gamma(\iota_{n-1})^{\pm 1} \ldots \Gamma(\iota_1)^{\pm 1}\, \Gamma(\iota_o)^{\pm 1}\, s(i) = g\, s(i).$

PROPOSITION 8.1. Let $\underline{A}$ be a group G. If (G, s) is a lower bound of a connected diagram $\Gamma : I \to \underline{A}$, then $\inf \Gamma = (G, s)$. If G has more than one element, a disconnected diagram $\Gamma : I \to \underline{A}$ has no inf.

<u>Proof.</u> First assume that Γ is a connected diagram. Given two indices i and j which are connected by a sequence (*), we then have $t(j) = g\, s(i)$, where g is the element of G determined by (**). Therefore $t(j) = s(j)h$, where $h = s(i)^{-1}\, t(i) \in G$. Clearly h is the only element of G such that $t(j) = s(j)h$, hence (G, s) is in fact the inf of Γ. (We regard i as fixed, j as any element of I.)

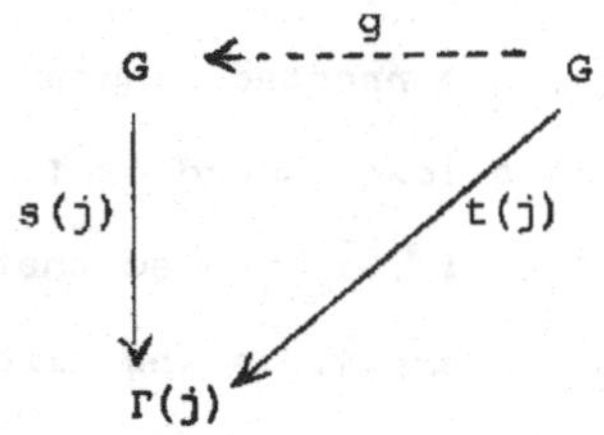

Next consider a disconnected diagram $\Gamma : I \to \underline{A}$. We may assume that I is the union of two nonempty categories I_1 and I_2 such that $[i_1, i_2]$ and $[i_2, i_1]$ are empty for all i_1 in I_1 and i_2 in I_2.

Let (G, s) be any lower bound of Γ. Take any element $h \neq 1$ of G and define a new lower bound (G, t) of Γ as follows:

$$t(i) = s(i) \quad \text{if } i \text{ is in } I_1$$
$$= s(i)h \quad \text{if } i \text{ is in } I_2 .$$

Clearly there is no element g of G such that $t(i) = s(i)g$ for all i in I. Therefore (G, s) is not the inf of Γ. This completes the proof.

PROPOSITION 8.2. If $\underline{A}$ is a group G, every functor $T : \underline{A} \to \mathrm{Ens}$ preserves infs.

<u>Proof.</u> If G has only one element 1 and $\Gamma : I \to G$, then $\inf \Gamma = (G, 1_I)$, where $1_I(i) = 1$ for all i in I. But then also $\inf (T\circ\Gamma) = \big(T(G), T\circ 1_I\big)$, as is easily verified.

Now assume that G has more than one element and that $\inf \Gamma = (G, s)$. Then I is connected, by Proposition 8.1. Surely $\big(T(G), T\circ s\big)$ is a lower bound of $T\circ\Gamma$, we will show that it is the inf.

Consider any other lower bound (X, u) of $T\circ\Gamma$. Again, let i be a fixed index, j any index, connected by the sequence (*). Put $f = T\big(s(i)^{-1}\big)\, u(i) : X \to T(G)$. Then, in view of (**),

$$T\big(s(j)\, f = T(g) T\big(s(i)\big) T\big(s(i)^{-1}\big) u(i) = T(g)\, u(i)$$
$$= T\big(\Gamma(\iota_{n-1})^{\pm 1}\big) \ldots T\big(\Gamma(\iota_o)^{\pm 1}\big)\, u(i).$$

Now, by naturality of u,

$$u(i_{k+1}) = T\big(\Gamma(\iota_k)^{\pm 1}\big)\, u(i_k),$$

for $k = 0, 1, \ldots, n-1$. Applying this repeatedly to the above, we obtain

$$T\big(s(j)\big)\, f = u(j).$$

Moreover, f is clearly unique with this property. Thus $\big(F(G),\ F\circ s\big)$ is indeed $\inf \Gamma$, and so F preserves infs, as to be shown.

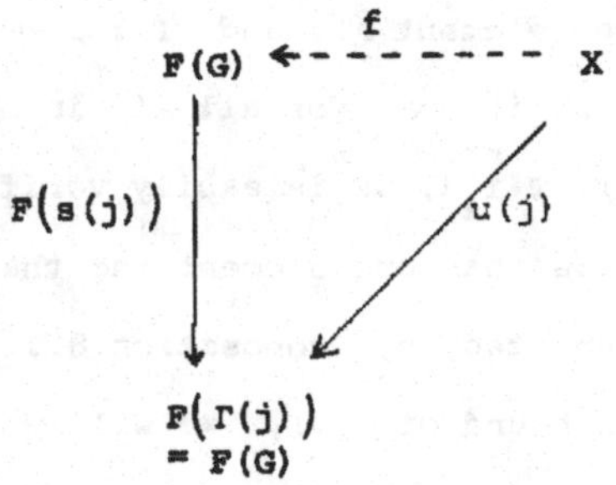

COROLLARY. If $\underline{A}$ is a group G, then $[\underline{A}^{o}, \text{Ens}]_{\inf} = [\underline{A}^{o}, \text{Ens}]$ is the category of all permutational representations of the opposite group.

9. Completions of categories of algebras. Let $\underline{C}$ be an equationally defined algebraic category with finitary operations, $\underline{A}$ a small subcategory of $\underline{C}$. Let $\underline{A}^*$ be the subcategory of $\underline{C}$ which consists of all algebras C such that $[-, C] : \underline{A}^o \to Ens$ preserves infs. In view of Proposition 3.2, $\underline{A}^*$ is the largest subcategory $\underline{B}$ of $\underline{C}$ containing $\underline{A}$ such that the inclusion functor $\underline{A} \to \underline{B}$ preserves sups.

We will show that, under fairly general circumstances, $[\underline{A}^o, Ens]_{inf}$ is equivalent to $\underline{A}^*$. If $\underline{C}$ is the category of R-modules, this result was first pointed out to the author by Ulmer. The proof in general will rely heavily on known results by Isbell and Lawvere.

PROPOSITION 9A (Isbell). Let $\underline{C}$ be an equationally defined algebraic category with operations at most n-ary. Let $\underline{F}$ be a small subcategory of $\underline{C}$ which contains the free algebra in n generators. Then $\underline{F}$ is a left adequate subcategory of $\underline{C}$.

For example, if $\underline{C}$ is the category of groups, the assumption is that $\underline{F}$ contains the free group in two generators, because multiplication is a binary operation. [See Isbell (1960), 2.2 and 1.1.]

PROPOSITION 9B (Lawvere). Let $\underline{C}$ be an equationally defined algebraic category with operations at most n-ary and equations involving at most n variables. Let $\underline{F}$ be the subcategory of $\underline{C}$ which consists of all free algebras in at most n generators. Then the functor $C \rightsquigarrow [-, C] : \underline{F}^{o} \to Ens$ is an equivalence of $\underline{C}$ with the category $[\underline{F}^{o}, Ens]_{prod}$ of all product preserving functors from $\underline{F}^{o}$ to Ens.

For example, if $\underline{C}$ is the category of groups, the assumption is that $\underline{F}$ consists of all free groups in at most three generators, because the associative law of multiplication involves three variables. [This result is implicit in Lawvere (1963) and (1965)].

PROPOSITION 9.1. Let $\underline{C}$ be an equationally defined algebraic category with operations at most n-ary and equations involving at most n variables. Let $\underline{A}$ be a small subcategory of $\underline{C}$ which contains all free algebras with at most n generators. Then $[\underline{A}^{o}, Ens]_{inf}$ is equivalent to $\underline{A}^{*}$.

Proof. Let $L : \underline{C} \to [\underline{F}^{o}, Ens]_{prod}$ be defined by $L(C) = [-, C] : \underline{F}^{o} \to Ens$. Proposition 9B asserts that there exists a functor $L' : [\underline{F}^{o}, Ens]_{prod} \to \underline{C}$ such that $L \circ L' \cong 1$ and $L' \circ L \cong 1$.

We now construct functors $M : \underline{A}^{*} \to [\underline{A}^{o}, Ens]_{inf}$ and $M' : [\underline{A}^{o}, Ens]_{inf} \to \underline{A}^{*}$. For any algebra B in $\underline{A}^{*}$ we write $M(B) = [-, B] : \underline{A}^{o} \to Ens$. If T is any functor in $[\underline{A}^{o}, Ens]_{inf}$, its restriction $T/_{\underline{F}}$ to $\underline{F}$ is an object of

$[\underline{F}^{o}, \text{Ens}]_{inf} \subset [\underline{F}^{o}, \text{Ens}]_{prod}$. We write $M'(T) = L'(T/_{\underline{F}})$. We claim that $M \circ M' \cong 1$ and $M' \circ M \cong 1$.

Indeed, take any B in $\underline{A}^*$, then

$$M'\big(M(B)\big) = L'\big(M(B)/_{\underline{F}}\big) = L'\big(L(B)\big) \cong B,$$

naturally in B. On the other hand, take any T in $[\underline{A}^{o}, \text{Ens}]_{inf}$, and put $T' = M\big(M'(T)\big)$. Then

$$T'/_{\underline{F}} = L\big(M'(T)\big) = L\big(L'(T/_{\underline{F}})\big) \cong T/_{\underline{F}} \ .$$

Now $T, T' : \underline{A}^{o} \to \text{Ens}$ preserve infs, and $\underline{F}^{o}$ is a right adequate subcategory of $\underline{A}^{o}$, by Proposition 9A. Therefore we may apply Lemma 5.3, or rather its dual, and obtain $T \cong T' = M\big(M'(T)\big)$. It may be verified that this isomorphism is natural in T, hence $M \circ M' \cong 1$. This completes the proof.

10. Completions of categories of modules. We shall write $\underline{M}_R$ for the category of all right R-modules, where R is an associative ring with 1. (All modules are understood to be unitary.) We shall consider subcategories $\underline{B}$ and $\underline{C}$ of $\underline{M}_R$ which satisfy the following three conditions:

I. If B is in $\underline{B}$ and C is in $\underline{C}$, then $[B, C] = 0$.

II. If $[B,C] = 0$ for all B in $\underline{B}$, then C is in $\underline{C}$.

III. If $[B,C] = 0$ for all C in $\underline{C}$, then B is in $\underline{B}$.

Well-known examples, when R is the ring of integers, are the following : $\underline{B}$ all torsion groups, $\underline{C}$ all torsion-free groups; $\underline{B}$ all divisible groups, $\underline{C}$ all reduced groups, i.e., all Abelian groups without nonzero divisible subgroups.

It is not difficult to show that $\underline{B}$ is closed under submodules if and only if $\underline{C}$ is closed under essential extensions, i.e., C in $\underline{C}$ and C' an essential extension of C implies C' in $\underline{C}$. In this case one might call the pair $\underline{B}$, $\underline{C}$ a "torsion theory". It follows from the work of Gabriel (1962) that a torsion theory is completely described by a certain set of right ideals of R. I am told that torsion theories have also been investigated by Spencer E. Dickson.

A pair of subcategories $\underline{B}$, $\underline{C}$ of $\underline{M}_R$ satisfying conditions I to III may be constructed from any subcategory $\underline{A}$ of $\underline{M}_R$: Let $\underline{B} = \underline{B}(\underline{A})$ consist of all modules B such that $[B, A] = 0$ for all A in $\underline{A}$, then let $\underline{C} = \underline{C}(\underline{A})$ consist of all modules C such that $[B, C] = 0$ for all B in $\underline{B}(\underline{A})$. For example, when R is the ring of integers and $\underline{A}$ consists only of the group of rationals, then $\underline{B}(\underline{A})$ is the category of torsion groups and $\underline{C}(\underline{A})$ is the category of torsion-free groups. We shall show that, under various conditions on $\underline{A}$, $\underline{C}(\underline{A}) = \underline{A}^*$, the subcategory of $\underline{M}_R$ which consists of all modules C such that the functor $[-, C] : \underline{A}^o \to Ens$ preserves infs. (See Section 9.)

We begin by listing some elementary properties of $\underline{B}$ and $\underline{C}$.

PROPOSITION 10.1. Let $\underline{B}$, $\underline{C}$ be a pair of subcategories $\underline{M}_R$ satisfying conditions I to III. Then the following statements are true:

(1) $\underline{B}$ is closed under factor modules and under sups.

(1*) $\underline{C}$ is closed under submodules and under infs.

(2) Every module M has a largest submodule βM in $\underline{B}$.

(2*) Every module M has a finest factor module $M/\gamma M$ in $\underline{C}$.

(3) M is in $\underline{B}$ if and only if it has no nonzero factor module in $\underline{C}$.

(3*) M is in $\underline{C}$ if and only if it has no nonzero submodule in $\underline{B}$.

(4) If B and M/B are in $\underline{B}$ then so is M.

(4*) If C and M/C are in $\underline{C}$ then so is M.

(5) $\beta M = \gamma M$.

Proof. Most of this is easy. We shall confine attention to statements (4) and (5).

(4) Suppose B and M/B are in $\underline{B}$. Let $m : B \to M$ be the inclusion mapping, and consider any mapping $f : M \to C$, where C is in $\underline{C}$. Then $fm : B \to C$ must be zero, hence $B \subset \mathrm{Ker} f$. Let $e : M \to M/B$ be the canonical epimorphism, then there exists $g : M/B \to C$ such that $ge = f$. Since M/B is in $\underline{B}$, $g = 0$, hence $f = 0$.

(5) On the one hand, the composite mapping $\beta M \to M \to M/\gamma M$ must be zero, hence $\beta M \subset \gamma M$. On the other hand, let B be in $\underline{B}$ and $f : B \to M/\beta M$. Then the image of f has the form $N/\beta M$, where $\beta M \subset N \subset M$. It is in $\underline{B}$, by (1). Hence N is in $\underline{B}$, by (4). Thus $N \subset \beta M$, by (3). Therefore $f = 0$, and so $M/\beta M$ is in $\underline{C}$. But $M/\gamma M$ is finer, by (2*), hence $\gamma M \subset \beta M$.

COROLLARY 1. Let $\underline{A}$ be a subcategory of $\underline{M}_R$ which is closed under submodules. Then a module M is in $\underline{B}(\underline{A})$ if and only if it has no nonzero factor module in $\underline{A}$, and a module M is in $\underline{C}(\underline{A})$ if and only if every nonzero submodule of M has a nonzero factor module in $\underline{A}$.

We omit the straight-forward proof.

COROLLARY 2. Let $\underline{B}$, $\underline{C}$ be a pair of subcategories of $\underline{M}_R$ satisfying conditions I to III. Then $\underline{C}$ is a left reflective subcategory and $\underline{B}$ is a right reflective subcategory.

Proof. We shall prove the statement concerning $\underline{C}$. Let $p : M \to M/\gamma M$ be the canonical epimorphism; we claim that it is a best approximation of M in $\underline{C}$. (See Section 3.)

Indeed, suppose $f : M \to C$, where C is in $\underline{C}$. Write $f = me$, m mono, e epi, with image C'. C' is in $\underline{C}$, by (1*). In view of (2*), there exists a unique mapping $x : M/\gamma M \to C'$ such that $e = xp$. Therefore $f = mxp$. Since p is epi, $m x$ is the unique y such that $f = yp$.

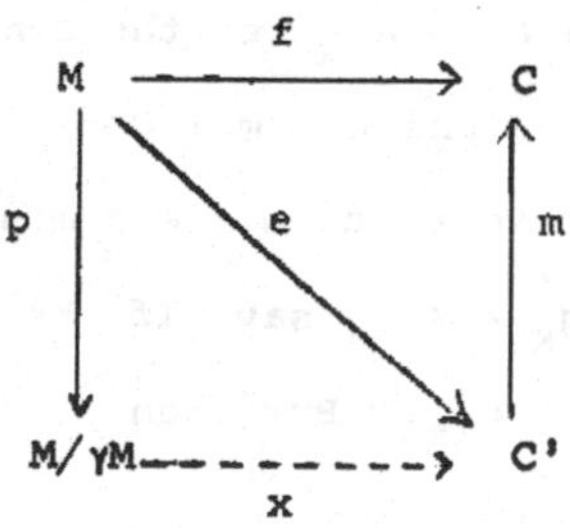

PROPOSITION 10.2. Let $\underline{A}$ be a subcategory of $\underline{M}_R$. Then any submodule of a product of modules from $\underline{A}$ is in $\underline{C}(\underline{A})$. The converse holds when $\underline{A}$ is closed under essential extensions.

Proof. First, consider a module M and assume that $m : M \to P = \Pi_{i \in I} A_i$ is mono, where the A_i are objects of $\underline{A}$. Let $p_i : P \to A_i$ be the canonical epimorphisms, and consider any homomorphism $f : B \to M$, where B is in $\underline{B}(\underline{A})$. Then each $p_i mf = 0$, as it sends B into A_i. Thus $mf = 0$, hence $f = 0$. Therefore M is in $\underline{C}(\underline{A})$.

Conversely, assume that $\underline{A}$ is closed under essential extensions. Consider any C in $\underline{C}(\underline{A})$ and any nonzero submodule K of C. By (3*), K is not in $\underline{B}(\underline{A})$. Therefore there exists a nonzero homomorphism $f_K : K \to A_K$, with A_K in $\underline{A}$. Extend f_K to $g_K : C \to A'_K$, where A'_K is a suitable essential extension of A_K, for instance the injective hull of A_K, hence also an object of $\underline{A}$. Put $P' = \Pi_K A'_K$, where K ranges over all nonzero submodules of C, and let $p'_K : P' \to A'_K$ be the canonical epimorphism. Then there exists a unique $m : C \to P'$ such that $p'_K m = g_K$. We shall prove that m is a monomorphism.

Indeed, $\text{Ker } m = \cap_K \text{ Ker } g_K = K^*$, say. If $K^* \neq 0$, this would be contained in $\text{Ker } g_{K^*}$. But then

$$0 = g_{K^*} K^* = f_{K^*} K^* \neq 0,$$

by choice of f_{K^*}, a contradiction. Therefore $\text{Ker } m = 0$, and our proof is complete.

EXAMPLE. Suppose $\underline{A}$ consists only of Q/Z or of all subgroups of Q/Z , where Q is the group of rationals and Z the group of integers. Then $\underline{C}(\underline{A})$ is the category of all Abelian groups.

LEMMA 10.1. Let $\underline{A}$ be a subcategory of $\underline{M}_R$ which is closed under submodules. Suppose the module M has a best approximation $x : M \to A$ in $\underline{A}$. Then x is epi.

The proof is routine and will be omitted.

LEMMA 10.2. Let $\underline{A}$ be a subcategory of $\underline{M}_R$ and assume that, whenever a module C in $\underline{C}(\underline{A})$ has a best approximation $x : C \to A$ in $\underline{A}$, x is an isomorphism. Then $\underline{C}(\underline{A}) \subset \underline{A}^*$.

<u>Proof.</u> In view of Proposition 3.2, it suffices to show that the inclusion $\underline{A} \to \underline{C}(\underline{A})$ preserves sups. Let $\Gamma : I \to \underline{A}$ with sup $\Gamma = (A, u)$ in $\underline{A}$. We claim this still holds in $\underline{C}(\underline{A})$.

Indeed, put sup $\Gamma = (M, v)$ in $\underline{M}_R$. Now we have a best approximation $e : M \to C = M/\gamma M$ of M in $\underline{C}(\underline{A})$. By a known theorem, sup $\Gamma = (C, e_I \circ v)$ in $\underline{C}$ (see Proposition 3.4.). Therefore there exists a unique $x : C \to A$ such that $x\, e\, v(i) = u(i)$ for all i in I. By Lemma 3.2, x is a best approximation of C in $\underline{A}$. By assumption, x is an isomorphism. Therefore sup $\Gamma = (A, u)$ in $\underline{A}$, as was to be shown.

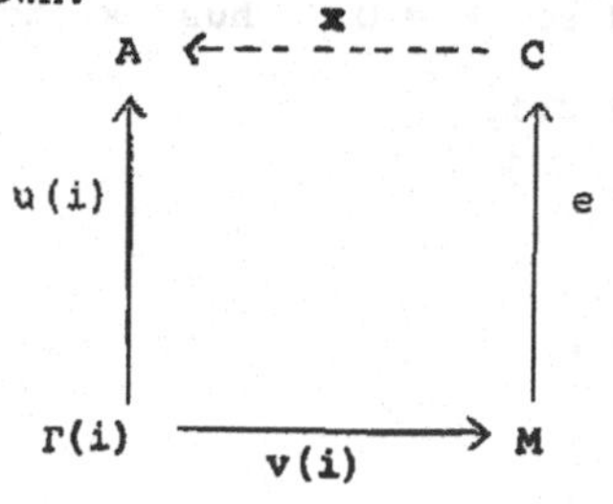

PROPOSITION 10.3. Let $\underline{A}$ be a subcategory of $\underline{M}_R$ which is closed under submodules. Then $\underline{C}(\underline{A}) \subset \underline{A}^*$ in any of the following three cases:

CASE 1. $\underline{A}$ is closed under essential extensions.

CASE 2. If A and M/A are in $\underline{A}$, so is M.

CASE 3. All modules in $\underline{A}$ are projective.

__Proof.__ We will show that the assumption of Lemma 10.2 is satisfied. Let C be any module in $\underline{C}(\underline{A})$ and $x : C \to A$ its best approximation in $\underline{A}$, assuming this to exist. Put $k : K \to C$ for the kernel of x. We shall see presently that in all three cases the following assertion holds:

(*) Every epimorphism $e : K \to A'$, with A' in $\underline{A}$, can be extended to a homomorphism $f : C \to A''$, where A'' is an extension of A' in A.

Suppose $e : K \to A'$ is an epimorphism with A' in $\underline{A}$. For the moment we assume (*). Since $x : C \to A$ was a best approximation in $\underline{A}$, there exists a unique $a : A \to A''$ such that $ax = f$. Write $k' : A' \to A''$ for the extension in (*), then $k'e = fk = axk = 0$. Since k' is mono, $e = 0$. Therefore K is in $\underline{B}(\underline{A})$, and so $K = 0$. Thus x is mono. Moreover, x is epi, by Lemma 7.1.

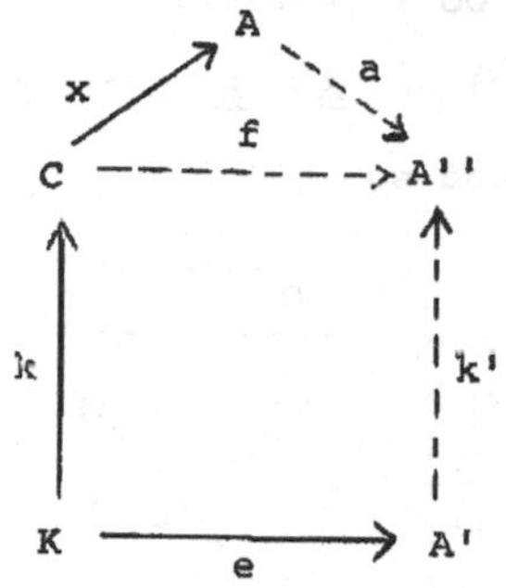

It remains to verify (*) in the three cases:

Case 1. Take A'' to be the injective hull of A.

Case 2. Let K' be the kernel of e, and take A'' = C/K', an extension of K/K' ≅ A'. Then K/K' and C/K ≅ (C/K')/(K/K') are in A, hence so is C/K'.

Case 3. In this case, K is a direct summand of C. Hence e may be extended to f, where A'' = A'.

This completes the proof.

LEMMA 10.3. Let $\underline{A}$ be a subcategory of $\underline{M}_R$ and assume that $\underline{A}$ is sup-dense in $\underline{A}^*$. Then $\underline{A}^*$ has only the zero module in common with $\underline{B}(\underline{A})$.

The assumption holds, for example, when $\underline{A}$ is small and contains R + R, in view of Proposition 9A.

<u>Proof.</u> Suppose B is common to $\underline{A}^*$ and $\underline{B}(\underline{A})$. By assumption, there exists a diagram $\Gamma : I \to \underline{A}$ such that $\sup \Gamma = (B, v)$ in $\underline{A}^*$. Suppose $t(i) : \Gamma(i) \to A$, naturally in i. Then there exists a unique $f : B \to A$ such that $fv(i) = t(i)$. However, since B is in $\underline{B}(\underline{A})$, $f = 0$, hence

$t(i) = 0$. We may write this $00(i) = t(i)$, where $0(i) = 0$. Thus $\sup \Gamma = (0, 0)$ in $\underline{A}$. Since $\underline{A} \to \underline{A}^*$ preserves sups, also $\sup \Gamma = (0, 0)$ in $\underline{A}^*$, and so $B = 0$.

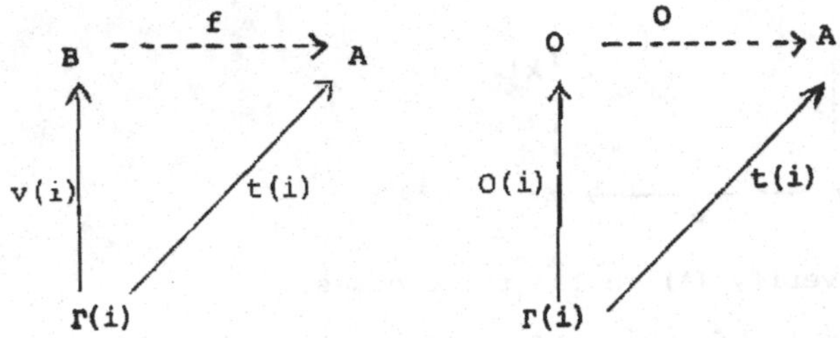

PROPOSITION 10.4. Let $\underline{A}$ be a subcategory of $\underline{M}_R$; and assume that $\underline{A}$ is sup-dense in $\underline{A}^*$ and that $\underline{C}(\underline{A}) \subset \underline{A}^*$. Then $\underline{C}(\underline{A}) = \underline{A}^*$.

<u>Proof.</u> Consider any module M in $\underline{A}^*$. Then $M/\beta M$ is in $\underline{C}(\underline{A})$, hence in $\underline{A}^*$. Now, by Proposition 3.2, $\underline{A}^*$ is closed under infs in $\underline{M}_R$. Moreover, βM may be regarded as the inf of a diagram $M \overset{p}{\underset{0}{\rightrightarrows}} M/\beta M$, where p is the canonical epimorphism and 0 the zero map. Therefore βM is in $\underline{A}^*$. By Lemma 10.3, $\beta M = 0$, hence M is in $\underline{C}(\underline{A})$. Thus also $\underline{A}^* \subset \underline{C}(\underline{A})$.

SUMMARY. We have shown that $\underline{A}^* = \underline{C}(\underline{A})$ under the following assumptions:

(1) $\underline{A}$ is sup-dense in $\underline{A}^*$.

(2) $\underline{A}$ is closed under submodules.

(3) One of the three cases of Proposition 10.3 holds.

In view of (2), we may then say that $\underline{A}^*$ consists of all those modules M such that every nonzero submodule of M has a nonzero factor module in $\underline{A}$. (See Corollary 1 to Proposition 10.1.)

We have already pointed out that (1) will hold if $\underline{A}$ is small and contains R + R. If even R + R + R is in $\underline{A}$, Proposition 9.1 allows us to conclude that $[\underline{A}^o, \mathrm{Ens}]_{inf}$ is equivalent to $\underline{C}(\underline{A})$.

Actually, when R + R is in $\underline{A}$, $\underline{C}(\underline{A})$ contains the left adequate subcategory {R + R}, by Proposition 9A and Lemma 6.1. Moreover, $\underline{C}(\underline{A})$ is closed under submodules and products, hence under direct sums. By Proposition 7.2, $\underline{C}(\underline{A})$ is sup-complete. Thus, when $\underline{A}$ is small, contains R + R + R, and satisfies (2) and (3), $[\underline{A}^o, \mathrm{Ens}]_{inf}$ will be sup-complete. We present a few examples of this last situation.

EXAMPLE 1. Let I be the injective hull of R (regarded as a right R-module) and suppose $\underline{A}$ consists of all submodules of I + I + I. (Case 1 of Proposition 10.3.)

EXAMPLE 2. Let R be the group of reals and Z the group of integers, and suppose $\underline{A}$ consists of all submodules of R/Z . (Observe that A contains a copy of Z + Z + Z , and again apply Case 1 of Proposition 10.3.)

EXAMPLE 3. Let R be right Noetherian, and suppose A consists of all Noetherian right R-modules. (Apply Case 2 of Proposition 10.3. We have stipulated that R is Noetherian to make sure that it is in A. In this case, "Noetherian" is the same as "finitely generated".)

EXAMPLE 4. Let R be right semihereditary , and suppose A consists of all finitely generated projective right R-modules. (Either Case 2 or Case 3 of Proposition 10.3 will apply.)

EXAMPLE 5. A consists of the Abelian groups
0, Z, Z + Z, Z + Z + Z.
(Case 3 of Proposition 10.3.)

POSTSCRIPT

Since these notes were typed there has appeared the following important paper on the same subject: John R.Isbell, Structure of categories, Bull.Amer.Math.Soc.,72(1966), 619-655. Here is a list of closely related results:

Lambek	Isbell
Corollary to Proposition 2.1	3.3c
Proposition 3.2	3.4
Proposition 6.1	3.10
Proposition 7.1	3.12
Lemma 7.1	2.4

REFERENCES

Bernhard Banaschewski, Hüllensysteme und Erweiterungen von Quasi- Ordnungen, Zeitschr. f. math. Logik und Grundlagen d. Math., 2(1956), 117-130.

Jean Benabou, Critères de représentabilité des foncteurs, C.R. Acad. Sc. Paris, 260 (1965), 752-755.

Beno Eckmann and Peter J. Hilton, Group-like structures in general categories, Math. Annalen, 145 (1962), 227-255; 150 (1963), 165-187; 151 (1963), 150-186.

Peter Freyd, Abelian categories, Harper and Row, New York 1964.

Pierre Gabriel, Des catégories abéliennes, Bull. Soc. Math. France, 90 (1962), 323-448.

Alexander Grothendieck, Sur quelques points d'algèbre homologique, Tôhoku Math. J., 9 (1957), 119-221.

J.R. Isbell, Adequate subcategories, Illinois J. Math., 4 (1960), 541-552.

.............., Subobjects, adequacy, completeness and categories of algebras, Rozprawy Matematyczne, 34 (1964), 1-33.

D. N. Kan, Adjoint functors, Trans. Amer. Math. Soc., 87 (1958), 294-329.

F. William Lawvere, Functorial semanties of algebraic theories, Proc. Nat. Acad. Sci., 50 (1963), 869-872.

.............., An elementary theory of the category of sets, Proc. Nat. Acad. Sci., 52 (1964), 1506-1511.

.............., Algebraic theories, algebraic categories, and algebraic functors, Symposium on the Theory of Models, North-Holland Publ. Co., Amsterdam 1965, 413-418.

.............., The category of categories as a foundation for mathematics. (Manuscript).

Jean-M. Maranda, Some remarks on limits in categories, Can. Math. Bull., 5 (1962), 133-146.

Saunders MacLane, Categorical algebra, Bull. Amer. Math. Soc., 71 (1965), 40-106.

Barry Mitchell, Theory of categories, Academic Press, New York 1965.

Friedrich Ulmer, Satelliten und derivierte Funktoren I, Math. Zeitschr., 91 (1966), 216-266.

McGill University, Montreal and Math. Research Institute, ETH, Zürich.

Offsetdruck: Julius Beltz, Weinheim/Bergstr

Lecture Notes in Mathematics

Bisher erschienen/Already published